Collective Phenomena in Plasmas and Elsewhere

# Collective Phenomena in Plasmas and Elsewhere

*Kinetic and Hydrodynamic Approaches*

Kamel Ourabah

iSTE

WILEY

First published 2023 in Great Britain and the United States by ISTE Ltd and John Wiley & Sons, Inc.



ISTE Ltd
27-37 St George's Road
London SW19 4EU
UK

www.iste.co.uk

John Wiley & Sons, Inc.
111 River Street
Hoboken, NJ 07030
USA

www.wiley.com



Library of Congress Control Number: 2023938460

British Library Cataloguing-in-Publication Data
A CIP record for this book is available from the British Library
ISBN 978-1-78630-769-9

# Contents

# Preface

*I am astounded by the number of people*
*who want to "know" the universe*
*when it's hard enough to find*
*your way around Chintatown.*

Woody Allen

During my physics studies, plasma physics classes always seemed boring to me. I could not say exactly why, but I aspired to fundamental things and struggled to see an essential aspect in plasma courses. It is therefore quite natural that I have turned to more fundamental branches of physics. It was only when I found myself joining a plasma physics laboratory, by a combination of circumstances, that I questioned my beliefs. I discovered that the tools and methods I associated with plasma physics were not confined to the study of plasmas but had a much more general scope.

Gradually, I discovered the exciting theories behind the study of plasmas in a clearer way, namely statistical physics and the kinetic theory of gases. Moreover, I also discovered, much to my surprise, the usefulness of methods that I thought were specific to plasmas in fields as diverse and varied as astrophysics, cosmology, dark matter models, the study of Bose–Einstein condensates, particle physics, etc. I reconciled myself, so to speak, with plasma physics to such an extent that I decided to write this book.

Beyond certain phenomena specific to plasmas, there are two major approaches in the study of plasmas and, more generally, in the study of complex systems, composed of a large number of degrees of freedom, namely the *hydrodynamic* approach and the *kinetic* approach. Each approach has its virtues and limitations. This book summarizes the theoretical bases of these two approaches and presents the qualities and limitations of each, as well as the bridges that exist between them. These two approaches have innumerable applications. Some of them are presented in this book; a drop of water in an ocean!

There are excellent books covering plasma physics in a much more comprehensive way. This book is perhaps less conventional in that it attempts to emphasize the more general character of the methods usually attributed (rightly or wrongly) to plasma physics. The methods are developed in a very general theoretical framework, followed by applications in plasmas, as well as in gravitational systems, the study of Bose–Einstein condensates, cosmology and dark matter models. The student I once was, who had no particular interest in plasma physics, would certainly have been delighted to open his eyes to the richness of these methods and the many fields in which they prove indispensable.

This book will hopefully be useful to master's and doctoral students, as well as young researchers, working in fields involving the study of large numbers of particles and collective phenomena, such as plasma theory, condensed matter theory, astrophysics and cosmology. As scientific research becomes more and more interdisciplinary, this book may be of increasing interest in the future. Indeed, it is essential today for a physicist to have an overview of the different branches of physics, and to have even somewhat superficial knowledge. It is also essential to know the bridges that exist between the different branches and to aspire to build new ones.

As the process of scientific research is itself a collective phenomenon, I have benefited from a lot of exchanges with colleagues. These exchanges contributed greatly to guiding my thinking and allowed me to crystallize these ideas in this book. In particular, exchanges with Hugo Terças and Tito Mendonça opened my eyes to the richness and flexibility of the quantum kinetic approach. Many colleagues and friends also encouraged me (sometimes almost to the point of threatening!) to write this book. I will not be able to name them all here, but they will know who they are.

June 2023

# Introduction

*All models are wrong,*
*but some are useful.*

George Box

The notion of an isolated system is a theoretical idealization. In reality, everything in the Universe is connected to everything and everything interacts with everything. It is indeed difficult to imagine a free particle, perfectly isolated from the rest of the Universe. And even if such a situation were conceivable, it would be quite uninteresting. On a practical level too, it is extremely difficult to perform an experiment on a small number of particles that would be, and remain throughout the experiment, isolated from the rest of the Universe. This would indeed imply absolute control of any form of interaction or exchange between the particle or particles in question and their immediate environment; it would be a very difficult task! Thus, even if the problem involves one or a small number of particles, an ideal model would take into consideration the exchanges of energy with the rest of the Universe, which would thus play the role of heat reservoir (and/or of particles). Such a model would contain, at least in principle, the dynamics of all the particles or degrees of freedom that constitute the environment.

A lesson to be drawn from this is that a model that is used for making reliable predictions, and is easily testable and empirically verifiable, must necessarily involve (at least indirectly) a large number of degrees of freedom. This is indeed the reason why statistical methods are ubiquitous in physics.

In recent decades, prodigious technical feats have been achieved in situ, that is to say experiments carried out in the laboratory. It is indeed possible today to make measurements of extreme precision and in increasingly high energy regimes. These experiments, however, come at a cost. Moreover, it may be advantageous to turn to the world's largest laboratory, that is, look at the sky and compare our theoretical predictions with space and astrophysical observations. What do we see when we look at the sky? Essentially, stars and large expanses of partially or totally ionized gas: this is called a plasma. This is the modern interest in plasma physics (in addition to the practical interest they arouse, given their application in various technologies). Plasmas are indeed an excellent field for testing theoretical predictions and comparing them with observations.

However, it is important not to see plasma physics as a corpus of methods and tools *born in plasma physics and applicable to plasmas only*; the reality is more complex! Indeed, the entire kinetic approach to plasma physics arises from the kinetic theory of gases, developed essentially by Boltzmann in the 19th century (Cohen and Thirring 2012), for classic neutral gases. Furthermore, methods born within the framework of plasma physics, such as the integral transformations introduced by Landau (1946), find new life through more contemporary applications, in cosmology for example (Baym et al. 2017; Moretti et al. 2020). To illustrate this, and without going into technical details, let us take the example of quantum plasma, which can be kinetically or hydrodynamically described by the Schrödinger equation coupled with the Poisson equation for the Coulomb field. Replace the Poisson equation of the Coulomb field with the Poisson equation of the gravitational field and you have a dark matter model (Bernal and Guzmán 2006; Ourabah 2020b) (i.e. *scalar field dark matter*)! Thus, it is important not to see these tools as specific to plasma physics, but to see their application in plasmas as a particular case, an approach whose scope is much more general. It is precisely this aspect that this book attempts to highlight.

An ideal, though impracticable, model when dealing with a complex system composed of a large number of particles such as a plasma would be to write and solve the equations of motion of all the particles. This would be equivalent to writing $N \sim 10^{23}$ partial differential equations and integrating them, which implies knowing the position and initial velocity of each particle. Such a method is clearly out of our reach. Thus, Laplace had imagined a supernatural intelligence (Laplace 1814), that is, *Laplace's demon* (see Figure I.1), which would know, with infinite precision, the position and speed of each particle and could thus predict the evolution of any system, even of the entire Universe: "[...] nothing would be uncertain and the future just like the past would be present before its eyes". But, we are not Laplace's demons and we have to deal with the information we have and our limitations to process such information. The best strategy, in practice, is the one that will "sort" between superfluous information and information essential to predict the evolution of the system. There are two main approaches to this end: kinetic theory and hydrodynamic models.

**Figure I.1.** *Illustration of Laplace's demon by Ricard Solé*

A first strategy is to omit all details on a microscopic scale. Thus, instead of considering the individual equations of motion of each of the particles composing the system, this strategy proposes to describe the global evolution of the density of particles in the medium. Such an approach is called *hydrodynamic* or *fluid*. Although simplistic in a sense, it makes it possible to describe, for example, the propagation of sound waves in media and explains, to a certain extent, the formation of stars from dust gases, that is, *Jeans instability* (see Chapter 6).

Another, perhaps more ambitious, approach consists of introducing individualism insofar as it introduces the notion of *distributional function*. Such an approach no longer treats the particles composing the system as being identical to each other, but makes it possible to describe the differences, in speed or energy, between them. This is called *kinetic theory*. It is important to understand, however, that no method is better than the other in absolute terms. While it is true that kinetic theory is more accurate than the fluid model, it is also more difficult to implement in numerical simulations. These simulations are very often essential to compare theoretical predictions with observational measurements.

This book presents the theoretical bases of these two approaches, the virtues and limits of each, as well as the bridges that exist between them. The evolution of the book is not chronological; thus, the kinetic theory will be presented before the hydrodynamic model, although historically, the fluid model was born before the kinetic theory. This choice is motivated by the fact that hydrodynamic equations can be demonstrated from kinetic theory, according to certain approximations.

The book is divided into two main parts. Part 1, aims to present the kinetic and hydrodynamic approaches in a very general theoretical framework. The models are developed so as to be applicable to different physical situations, which can be treated on an equal footing. Part 2, on the other hand, brings together applications of these methods in specific physical situations, in plasmas, gravitational systems, Bose–Einstein condensates, as well as in cosmology and the study of dark matter.

More specifically, Chapter 1 will be devoted to kinetic theory and Chapter 2 is devoted to the hydrodynamic approach, both being developed in a classical, non-quantum, and non-relativistic framework. Chapter 3 presents a generalization of these methods within the framework of quantum mechanics. It aims to present the quantum hydrodynamic model and the quantum kinetic theory of Wigner–Moyal. These two approaches are indispensable tools for modern physics and allow for describing, on an equal footing, a large number of systems (Mendonça and Terças 2013; Mendonça 2019; Ourabah 2021), ranging from quantum plasmas to Bose–Einstein condensates and assemblies of cold atoms in optical networks. Chapter 4 presents a generalization of the kinetic and hydrodynamic approaches to the theory of relativity, restricted and general. Such generalization makes it possible to adapt hydrodynamic and kinetic methods to extreme astrophysical conditions, where the speeds of the particles approach the speed of light, and to curved spaces.

Part 2 brings together applications of kinetic and hydrodynamic approaches, and their generalizations to quantum mechanics and the theory of relativity, in various physical systems. When deemed necessary, both approaches will be applied to the same problem, allowing virtues and limitations of each to be seen. In other cases, one approach will be preferred, for simplicity, but with sufficient evidence in the use of the other method.

Chapter 5 will be devoted to the study of plasmas. We will mainly deal with oscillations in plasmas, both classical and quantum or relativistic. We will also study the phenomenon of *Landau damping*: a purely kinetic phenomenon that cannot be studied in a hydrodynamic framework. In Chapter 6, we will use the analogy between Colombian and gravitational interactions to study gravitational systems or, more precisely, systems of self-gravitating particles, that is, systems of particles interacting with each other via gravitational interactions. We will essentially study the phenomenon analogous to oscillations in plasmas, namely Jeans gravitational instability; it is the phenomenon that explains the formation of stars from molecular clouds. We will present different variants of the problem, for example, in the presence of dark matter and in the framework of alternative theories of gravity.

Chapter 7 will be devoted to the use of hydrodynamic and kinetic approaches to the study of Bose–Einstein condensates (a particular state of a quantum gas at very low temperature). Since Bose–Einstein condensates are quantum, we will present the use of quantum generalizations, namely the hydrodynamic-quantum model and the Wigner–Moyal kinetic theory. We shall see that, again, a phenomenon similar to the oscillations in plasmas exists in the Bose–Einstein condensates; this is what is called the *Bogoliubov excitations*. Finally, in Chapter 8, we will present applications of hydrodynamic and kinetic methods in the context of cosmology and the study of dark matter. We will mainly insist on the formulation of hydrodynamic equations in an expanding universe. We will also study a dark matter model, composed of scalar fields.

The list of applications presented here is, of course, far from exhaustive. The hydrodynamic and kinetic approaches are, in fact, highly flexible and find applications in an increasing number of fields. The applications discussed in this book can, however, be seen as a starting point for studying other systems that can be formally described by similar equations. Also, we will direct the reader, through a rich bibliography covering the state of the art of research in this field, to other applications in very active fields of scientific research.

# PART 1

# Theory

# 1

# Kinetic Theory

*The general struggle for existence of animate beings is not a struggle for raw materials [...], all abundantly available – nor for energy, which exists in plenty in any body in the form of heat Q, but of a struggle for entropy S, which becomes available through the transition of energy from the hot sun to the cold earth.*

Ludwig Boltzmann

A complex system, such as a plasma or a system of self-gravitating particles, shows collective behavior. Thus, the evolution of one particle strongly depends on the evolution of the other particles in the medium. This collective behavior is actually the consequence of multiple interactions between a large number of particles and the fields they generate (electric and magnetic fields for a plasma or the gravitational field for systems of self-gravitating particles). Thus, the movement of each particle has the consequence of modifying fields, which in turn influences the trajectory of all particles, thus generating a modification of fields, and so on.

Given the large number of particles making up the system, a dynamic description (i.e. based on the individual equations of motion of each particle) is clearly out of reach. Fortunately, such an impractical approach is not necessarily desirable. Indeed, it is often useless to know the individual behavior of each particle. It is more useful to know the statistical behavior of the system. The most precise statistical approach is theoretical kinetics, to which this chapter is devoted.

## 1.1. The impossibility of an accurate approach

As explained earlier, an accurate approach, based on the dynamics of each particle, is an insoluble problem. However, let us try to develop such a model. This will make it possible, firstly, to better see why such an approach is feasible in practice and, secondly, it will constitute a good starting point for building the foundations of kinetic theory.

A dynamic approach first requires defining the trajectories of each particle that make up the system. Thus, it is interesting to consider the position and velocity of each particle as *independent variables* a six-dimensional space. This is called the *phase space*. Note $\vec{x}$ the position of the particle and $\vec{v}$ its velocity vector. Thus, a particle is characterized, at a given instant $t$, by a point located in a volume $d^3xd^3v$ in the phase space. Thus, the evolution of the particle (whose position and velocity vary over time) corresponds to a trajectory in this six-dimensional space.

To simplify, let us first consider the *simple* case of a single point particle whose position in phase space can be described by a Dirac[1]. In real space (that of positions), this particle has a trajectory $\vec{\mathbf{X}}_1(t)$, representing the positions $\vec{x}$ occupied by the particle over time. Similarly, the particle has a trajectory $\vec{\mathbf{V}}_1(t)$ in velocity space. By combining the coordinates of space $\vec{x}$ and velocity $\vec{v}$, in a six-dimensional space, we can thus write the density of a particle as follows (Savoini 2010):

$$\mathcal{N}_1(\vec{x},\vec{v},t) = \delta\left[\vec{x} - \vec{\mathbf{X}}_1(t)\right]\delta\left[\vec{v} - \vec{\mathbf{V}}_1(t)\right] \quad [1.5]$$

1 Dirac's delta "function" is defined as

$$\delta(x-x_0) = \begin{cases} 0, & x \neq x_0 \\ +\infty, & x = x_0 \end{cases} \quad [1.1]$$

such that

$$\int_{-\infty}^{\infty} \delta(x-x_0)dx = 1 \quad [1.2]$$

It can be shown that for any continuous function $f(x)$, we have

$$\int_{-\infty}^{\infty} \delta(x-x_0)f(x)dx = f(x_0) \quad [1.3]$$

In the case of a vector $\vec{r} = \{x,y,z\}$, we can note that

$$\delta(\vec{r}) \equiv \delta(x)\delta(y)\delta(z) \quad [1.4]$$

This equation simply translates the fact that the density of the particle is 1 where the particle is located, in phase space, and is 0 everywhere else. When time varies, this density varies in turn under the effect of the forces to which the particle is subjected.

A system composed of a single particle is not very interesting but the previous reasoning can be easily extended to the case of $N$ particles. Indeed, for a system composed of $N$ particles (assumed to be of the same species, for example, $N$ identical molecules), the density is the sum of the individual densities of the set of particles, that is

$$\mathcal{N}(\vec{x},\vec{v},t)=\sum_{i=1}^{N}\delta\left[\vec{x}-\vec{X}_i(t)\right]\delta\left[\vec{v}-\vec{V}_i(t)\right] \quad [1.6]$$

Note that in the case of a system composed of two or more different species, as in the case of a plasma, equation [1.6] describes the density for each species of particles (electrons and ions, separately).

This density represents the volume occupied by all the particles in the phase space. As the number of particles is conserved, this volume can change shape during evolution, but its size remains the same. Thus, the total derivative of the density with respect to time, along the trajectory, must be zero, that is

$$\frac{d\mathcal{N}(\vec{x},\vec{v},t)}{dt}=0 \quad [1.7]$$

Let us now recall that the trajectory of a particle is determined by its position, which varies over time, under the effect of *microscopic*[2] fields to which the particle is subjected. Typically, this will be the electric field $\vec{\mathbf{E}}_m$ and magnetic field $\vec{\mathbf{B}}_m$ in the case of a plasma, or the gravitational field $\vec{\mathbf{g}}$, in the case of a self-gravitating particle system. For a particle $i$ of mass $m$, we then have:

$$\begin{aligned}\frac{d\vec{\mathbf{X}}_i(t)}{dt}&=\vec{\mathbf{V}}_i(t)\\ \frac{d\vec{\mathbf{V}}_i(t)}{dt}&=\frac{\vec{\mathbf{F}}}{m}\end{aligned} \quad [1.8]$$

where $\vec{\mathbf{F}}$ represents the force. In the case of a plasma, the dominating force will be the Lorentz force:

$$\vec{\mathbf{F}}=q\left[\vec{\mathbf{E}}_m\left(\vec{\mathbf{X}}_i(t),t\right)+\vec{\mathbf{V}}_i(t)\times\vec{\mathbf{B}}_m\left(\vec{\mathbf{X}}_i(t),t\right)\right] \quad [1.9]$$

2 We speak of *microscopic* as opposed to *external* fields, because these fields are generated by all the particles in the medium.

where $q$ is the charge of the particle. Under certain conditions, the magnetic field can be neglected and only the Coulomb force $q\vec{\mathbf{E}}_m$ will be retained. On the other hand, it will be the gravitational force for a gravitational system:

$$\vec{\mathbf{F}} = m\vec{\mathbf{g}}\left(\vec{\mathbf{X}}_i(t),t\right) \qquad [1.10]$$

It should be noted that microscopic fields are not external fields (and this is indeed the whole issue) but are produced in a self-consistent way, that is to say by the particles of the medium themselves. Thus, in the case of a gravitational system, in the Newtonian approximation, the gravitational field is simply "sourced" by the density $\rho_m$ (mass per unit volume).

So, we get:

$$\vec{\nabla}\cdot\vec{g}(\vec{\mathbf{x}},t) = -4\pi G\rho_m \qquad [1.11]$$

where $G$ is the gravitational constant, while in the case of a plasma, the electric $\vec{E_m}$ and magnetic $\vec{B_m}$ fields obey Maxwell's equations:

$$\begin{aligned}
&\vec{\nabla}\cdot\vec{E_m}(\vec{\mathbf{x}},t) = \frac{\rho_m(\mathbf{x},t)}{\varepsilon_o}\\
&\vec{\nabla}\cdot\vec{B_m}(\vec{\mathbf{x}},t) = 0\\
&\vec{\nabla}\times\vec{E_m}(\vec{\mathbf{x}},t) = -\frac{\partial\vec{B_m}(\vec{\mathbf{x}},t)}{\partial t}\\
&\vec{\nabla}\times\vec{B_m}(\vec{\mathbf{x}},t) = \mu_o\vec{j_m}(\vec{\mathbf{x}},t) + \mu_o\varepsilon_o\frac{\partial\vec{E_m}(\vec{\mathbf{x}},t)}{\partial t}
\end{aligned} \qquad [1.12]$$

where $\varepsilon_0$ represents the dielectric permittivity of the vacuum and $\mu_0$ is the magnetic permeability of the vacuum. The charge density $\rho$ (charge per unit volume) and the current $\vec{j_m}$ are defined by:

$$\begin{aligned}
&\rho_m(\vec{\mathrm{x}},t) = \sum_{s=e,i} q_s\int d\vec{\mathrm{v}}\,\mathcal{N}_s(\vec{\mathrm{x}},\vec{\mathrm{v}},t)\\
&\vec{j_m}(\vec{\mathrm{x}},t) = \sum_{s=e,i} q_s\int d\vec{\mathrm{v}}\,\vec{v}\,\mathcal{N}_s(\vec{\mathrm{x}},\vec{\mathrm{v}},t)
\end{aligned} \qquad [1.13]$$

the indices $e$ and $i$, respectively, representing the electrons and the ions. It can happen that other species are present in the plasma, for example, dust grains (which are usually negatively charged). In this case, the contribution of these additional species will have to be added to the previous equations.

We note that, so far, no approximation has been made. Indeed, in the case of a plasma for example, Maxwell's equations [1.12] determine the fields exactly. In addition, equation [1.9] determine the exact trajectories of particles, depending on these fields. We can also observe that this system of equations, namely the relationships [1.9] and [1.12], where the relationships [1.9] and [1.11] constitute a closed system, that is, if we know the position and the speed of each particle as well as the value of the microscopic fields at a given instant, we can *in principle* determine, in an exact way, the value of the positions, velocities and fields at any instant.

To better understand this, let us take equation [1.7] and explain the total derivative $d/dt$. So, we get:

$$\frac{d\mathcal{N}(\vec{x},\vec{v},t)}{dt}=\frac{\partial\mathcal{N}}{\partial t}+\frac{d\vec{x}}{dt}\cdot\frac{\partial\mathcal{N}}{\partial\vec{x}}+\frac{d\vec{v}}{dt}\cdot\frac{\partial\mathcal{N}}{\partial\vec{v}}=0 \qquad [1.14]$$

where the first term represents the direct dependence on time, while the remaining two terms represent the indirect dependence on time through position and velocity.

Expanding on all of the terms and using the definition of density [1.6], as well as the equations of motion [1.8], we can write:

$$\begin{aligned}\frac{d\mathcal{N}(\vec{x},\vec{v},t)}{dt}&=\frac{d}{dt}\sum_{i=1}^{N}\mathcal{N}_i=\sum_{i=1}^{N}\frac{d\mathcal{N}_i}{dt}=\frac{\partial}{\partial t}\left(\sum_{i=1}^{N}\delta\left[\vec{x}-\vec{X}_i(t)\right]\delta\left[\vec{v}-\vec{V}_i(t)\right]\right)\\&+\sum_{i=1}^{N}\vec{V}_i(t)\cdot\frac{\partial}{\partial\vec{X}_i}\left(\delta\left[\vec{x}-\vec{X}_i(t)\right]\delta\left[\vec{v}-\vec{V}_i(t)\right]\right)\\&+\sum_{i=1}^{N}\frac{\vec{F}}{m}\cdot\frac{\partial}{\partial\vec{V}_i}\left(\delta\left[\vec{x}-\vec{X}_i(t)\right]\delta\left[\vec{v}-\vec{V}_i(t)\right]\right)=0\end{aligned} \qquad [1.15]$$

This equation can be simplified, using the properties of Dirac delta distribution, by replacing $\vec{X}_i$ by $\vec{x}$ and $\vec{V}_i$ by $\vec{v}$. Thus, we get:

$$\begin{aligned}&\frac{\partial}{\partial t}\left(\sum_{i=1}^{N}\delta\left[\vec{x}-\vec{X}_i(t)\right]\delta\left[\vec{v}-\vec{V}_i(t)\right]\right)\\&+\vec{v}\cdot\frac{\partial}{\partial\vec{x}}\left(\sum_{i=1}^{N}\delta\left[\vec{x}-\vec{X}_i(t)\right]\delta\left[\vec{v}-\vec{V}_i(t)\right]\right)\\&+\frac{\vec{F}}{m}\cdot\frac{\partial}{\partial\vec{v}}\left(\sum_{i=1}^{N}\delta\left[\vec{x}-\vec{X}_i(t)\right]\delta\left[\vec{v}-\vec{V}_i(t)\right]\right)=0\end{aligned} \qquad [1.16]$$

which can be rewritten in a more compact form, since we note that sums no longer intervene as does density. So, we have[3]:

$$\frac{\partial \mathcal{N}}{\partial t} + \vec{v} \cdot \vec{\nabla}_x \mathcal{N} + \vec{a} \cdot \vec{\nabla}_v \mathcal{N} = 0 \qquad [1.18]$$

where $\vec{a} := \vec{\mathbf{F}}/m$ is the acceleration of the particle. Equation [1.18] is called the *Klimontovich equation*. Once coupled with Maxwell's equations, it allows an *exact* description of plasmas. Similarly, once coupled with equation [1.10] describing the variation of the gravitational field, it allows an exact description of a system of particles interacting through gravitational forces.

An exact expression you say? But the equation does not seem so complicated! In fact, the apparent simplicity of equation [1.18] is misleading because all difficulty is hidden behind the acceleration term $\vec{a}$ (or force $\vec{\mathbf{F}}$). In fact, this term depends on the microscopic fields as seen by each particle (Savoini 2010). If we wanted to calculate it exactly, its calculation would require the knowledge of exact trajectories of each of the particles, a task impossible in practice! Moreover, the density [1.6] is a very fluctuating quantity, because it is zero everywhere between the particles and is infinite (Dirac delta) for point particles. Microscopic fields (electric and magnetic fields or the gravitational field) are also very fluctuating, because they result from the density of particles, which itself fluctuates enormously in time and space.

Thus, the model described by the Klimontovich equation [1.18] is not of any use in practice. However, it is interesting to see it as a starting point for building a necessarily less precise statistical approach. Such an approach is, in fact, sufficient because it is not necessary to know the position and velocity of each particle, in real time, nor the exact value of fields. What interests us most often are the *mean fields* (generated by all the particles) which interact with a large number of particles, which can be treated statistically. This is precisely what the Boltzmann equation allows us to do.

## 1.2. Boltzmann equation

A statistical description is an "averaged" description in one sense, that is to say a description involving only macroscopic quantities that vary in time and space. Take,

3 We have defined the operators:

$$\begin{aligned} \vec{\nabla}_x &:= \frac{\partial}{\partial \vec{x}} = \frac{\partial}{\partial x}\vec{i} + \frac{\partial}{\partial y}\vec{j} + \frac{\partial}{\partial z}\vec{k} \\ \vec{\nabla}_v &:= \frac{\partial}{\partial \vec{v}} = \frac{\partial}{\partial v_x}\vec{i} + \frac{\partial}{\partial v_y}\vec{j} + \frac{\partial}{\partial v_z}\vec{k} \end{aligned} \qquad [1.17]$$

for example, the electric field; it may be useful to break it down into an average component $\overrightarrow{\langle E\rangle}$ (generated by all the particles in the medium) and a fluctuating part $\overrightarrow{\delta E}$ whose effect is only significant for particles, which would find themselves very close to each other. Also, in a statistical description, it is not necessary to know the exact position of each particle in phase space (an impossible task anyway!). It is possible, and even more useful, to know the number of particles in a given volume $\Delta\vec{x}\Delta\vec{v}$ of the phase space. It is possible, in fact, to count the number of particles located in a volume in phase space, at an instant $t$, having a velocity between $\vec{v}$ and $\vec{v}+\Delta\vec{v}$. Let us take this number, divide it by the size of this volume and multiply it by $\Delta\vec{v}$, and note the function, thus obtained $f(\vec{x},\vec{v},t)$. Thus built, this function represents the density of particles in the phase space. Thus, by integrating over the entire velocity space, we obtain the average density (average number of particles per unit volume):

$$n(\vec{x},t)=\int_{-\infty}^{+\infty} f(\vec{x},\vec{v},t)d^3\vec{v} \tag{1.19}$$

Similarly, by also summing over the positions, we have the total number of particles $N$, namely:

$$N=\int_{-\infty}^{+\infty}\int_{-\infty}^{+\infty} f(\vec{x},\vec{v},t)d^3\vec{v}d^3\vec{x} \tag{1.20}$$

The function $f(\vec{x},\vec{v},t)$ is called the *distribution function*. If we agree to start from an exact approach, we can, from the Klimontovich equation, obtain an equation describing the variation of the distribution function $f(\vec{x},\vec{v},t)$. Such an equation is, in fact, much more useful in practice.

Assuming that all of the quantities of densities and fields can be decomposed into an average component and a fluctuating component, we then have:

$$\begin{aligned}
&\mathcal{N}(\vec{x},\vec{v},t)=\langle\mathcal{N}(\vec{x}\vec{v},t)\rangle+\delta\mathcal{N}(\vec{x},\vec{v},t)=f(\vec{x}\vec{v},t)+\delta\mathcal{N}(\vec{x},\vec{v},t)\\
&\vec{\mathbf{g}}(\vec{x},t)=\vec{\mathrm{g}}(\vec{x},t)+\delta\vec{\mathrm{g}}(\vec{x},t)\\
&\vec{\mathbf{E}}_m(\vec{x},t)=\vec{\mathrm{E}}(\vec{x},t)+\delta\vec{\mathrm{E}}(\vec{x},t)\\
&\vec{\mathbf{B}}_m(\vec{x},t)=\vec{\mathrm{B}}(\vec{x},t)+\delta\vec{\mathrm{B}}(\vec{x},t)
\end{aligned} \tag{1.21}$$

where $<\vec{\mathbf{g}}>\equiv\vec{\mathbf{g}}, <\overrightarrow{\mathbf{E}_m}>\equiv\vec{\mathbf{E}}, <\overrightarrow{\mathbf{B}_m}>\equiv\vec{\mathbf{B}}$ are the mean fields and $<\delta\vec{\mathbf{g}}>=<\delta\vec{\mathbf{E}}>=<\delta\vec{\mathbf{B}}>=\langle\delta\mathcal{N}\rangle=0$. Using this definition and averaging the Klimontovich equation [1.18], we obtain, in the case of electric and magnetic fields:

$$\begin{aligned}&\frac{\partial f(\vec{x},\vec{v},t)}{\partial t}+\vec{v}\cdot\vec{\nabla}_x f(\vec{x},\vec{v},t)+\frac{q}{m}[\vec{E}(\vec{x},t)+\vec{v}\times\vec{B}(\vec{x},t)]\cdot\vec{\nabla}_v f(\vec{x},\vec{v},t)=\\&-\frac{q}{m}\left\langle[\delta\vec{\mathbf{E}}(\vec{x},t)+\vec{v}\times\overrightarrow{\delta B}(\vec{x},t)]\cdot\vec{\nabla}_v\delta\mathcal{N}(\vec{x},\vec{v},t)\right\rangle\end{aligned}\qquad[1.22]$$

A similar equation can be established in the case of a gravitational field, replacing the Lorentz force with the force of gravity $m\vec{g}$:

$$\begin{aligned}&\frac{\partial f(\vec{x},\vec{v},t)}{\partial t}+\vec{v}\cdot\vec{\nabla}_x f(\vec{x},\vec{v},t)+\vec{g}(\vec{x},t)\cdot\vec{\nabla}_v f(\vec{x},\vec{v},t)=\\&-\left\langle\delta\vec{\mathbf{g}}(\vec{x},t)\cdot\vec{\nabla}_v\delta\mathcal{N}(\vec{x},\vec{v},t)\right\rangle\end{aligned}\qquad[1.23]$$

It is interesting to note that the left-hand side in equations [1.22] and [1.23] contains only average values; these do not depend on the exact positions of particles in the phase space but simply on the coordinates $(\vec{x},\vec{v},t)$ of the phase space. Thus, the distribution $f(\vec{x},\vec{v},t)$ and the equation governing its evolution do not provide information on the exact position of each particle in a volume of phase space but rather on the probability that these particles find themselves there. The function $f(\vec{x},\vec{v},t)$ is therefore indeed a distribution function.

Equations [1.22] and [1.23] describe the variation of the distribution $f(\vec{x},\vec{v},t)$ under the action of mean fields.

It is also important to note that it is the right-hand side that contains all the dependencies on the fluctuating quantities $\delta\mathcal{N}, \delta\vec{\mathbf{g}}$ or $\delta\vec{\mathbf{E}}$ and $\delta\vec{\mathbf{B}}$, which represent all the short-range interactions and collisions[4] in the system. Its determination poses a very serious problem because the distribution $f(\vec{x},\vec{v},t)$ is a statistical quantity, which does not distinguish between particles and does not determine their exact positions in phase space. However, collisions are *individual* processes (involving two particles), and their determination requires, in principle, a microscopic description. A way to simplify this equation consists of neglecting the correlations between the fields and only taking into account the correlation between the particles, so as to attribute the right-hand side of equations [1.22] and [1.23] only to collisions.

4 Collisions occur when two particles, moving in a rectilinear motion, change direction, before resuming a rectilinear motion.

The equation is rewritten as follows:

$$\frac{\partial f(\vec{x},\vec{v},t)}{\partial t}+\vec{v}\cdot\vec{\nabla}_x f(\vec{x},\vec{v},t)+\frac{\vec{\mathbf{F}}}{m}\cdot\vec{\nabla}_v f(\vec{x},\vec{v},t)=\left(\frac{\delta f}{\delta t}\right)_{\text{coll}} \qquad [1.24]$$

where the right-hand side describes the temporal variation of the distribution $f(\vec{x},\vec{v},t)$ under the effect of collisions. Equation [1.24] is called the *Boltzmann equation* (established by Boltzmann in 1872). It is the basis of kinetic theory and is of remarkable use in the study of complex systems and collective phenomena. Its use, however, is quite subtle due to the collision term, which (for now) is unspecified and whose exact expression represents a serious mathematical challenge.

## 1.3. Collision term

Thus, the whole difficulty in using the Boltzmann equation [1.24] consists of knowing (at least in an approximate manner) the collision term. There are several methods to model the collision term, such as the Bhatnagar–Gross–Krook relaxation model, the Boltzmann collision integral, and the Fokker–Planck collision term. We will discuss these different approaches here.

### 1.3.1. *Absence of collisions: Vlasov equation*

The first approximation that will interest us is undoubtedly, because of its simplicity, the most used in kinetic theory. It simply consists of neglecting the effect of collisions.

Thus, the Boltzmann equation becomes:

$$\frac{\partial f(\vec{x},\vec{v},t)}{\partial t}+\vec{v}\cdot\vec{\nabla}_x f(\vec{x},\vec{v},t)+\frac{\vec{\mathbf{F}}}{m}\cdot\vec{\nabla}_v f(\vec{x},\vec{v},t)=0 \qquad [1.25]$$

The above equation is called the *Vlasov equation* or the collisionless Boltzmann equation. It was established, in the case of a plasma, by Vlasov in 1938 (Vlasov 1938).

In fact, the Vlasov equation [1.25] ignores all collisions characterized by local interactions, that is to say when two particles are very close to each other (where the fluctuating fields intervene). The only interactions, retained in the Vlasov equation, are actions *at a distance*, that is to say those related to the movement of a particle in the *mean field* generated by the set of other particles.

These fields depend on the position of all the particles in the medium and thus depend on the distribution $f(\vec{x}, \vec{v}, t)$: this is the reason why the Vlasov equation is nonlinear, which makes its analytical resolution is generally impossible. There are, however, so-called perturbative techniques, which make it possible to derive approximate solutions (see, for example, Chapter 5).

To better understand the physical meaning of the Vlasov equation [1.25] and its fundamental aspect, note that it can be rewritten in the following way:

$$\frac{Df}{Dt} = \frac{\partial f}{\partial t} + \vec{\nabla}_x(\vec{v} f) + \vec{\nabla}_v(\vec{a} f) = 0 \qquad [1.26]$$

where $D/Dt := \partial/\partial t + \vec{\nabla}x + \vec{\nabla}v$ represents what is called the *convective derivative*. The physical meaning of equation [1.26] is clearer: in the absence of collisions, the distribution function $f(\vec{x}, \vec{v}, t)$ does not vary over time along a trajectory in phase space. This can be seen as a link with a theorem of paramount importance in classical physics, namely *Liouville's theorem*. This theorem states that a volume element in phase space can deform during the evolution of the system but remains the same size. It is then said that the phase space behaves like an *incompressible* fluid.

### 1.3.2. *Bhatnagar–Gross–Krook relaxation model*

Another approximation makes it possible to introduce collisions in a remarkably simple way. This is based on the following observation: when the constraints applied to an nonequilibrium system cease, the system gradually returns to a state of equilibrium; this is called *relaxation*. However, it is observed experimentally that the temporal evolution of any distribution function $f(\vec{r}, \vec{v}, t)$ toward an equilibrium function[5] $f_{eq}(\vec{v})$ generally follows an exponential distribution, that is,

$$f - f_{eq} \propto \exp(-t/\tau) \qquad [1.27]$$

where $\tau$ is called *relaxation time*. It is inversely proportional to the frequency of collisions.

The law [1.27] amounts to writing:

$$\frac{\partial f}{\partial t} = \frac{f_{eq} - f}{\tau} \qquad [1.28]$$

5 The equilibrium state is a steady state, so the equilibrium distribution $f_{eq}(\vec{v})$ is not time dependent. It may, however, depend on the position $\vec{r}$ the presence of potential.

or:

$$\frac{\partial f(\vec{r},\vec{v},t)}{\partial t}+\frac{f(\vec{r},\vec{v},t)}{\tau}-\frac{f_{eq}(\vec{r},\vec{v})}{\tau}=0 \quad [1.29]$$

whose homogeneous solution is indeed exponential, namely:

$$f(\vec{v},t)=f_{eq}(\vec{v})+[f(\vec{v},0)-f_{eq}(\vec{v})]\exp\left(\frac{-t}{\tau}\right) \quad [1.30]$$

However, the main effect of collisions is to relax the distribution function toward an equilibrium distribution function. This is because particles exchange energy and momentum through collisions, which gradually bring the system into equilibrium. Thus, the collisions term can be written as:

$$\left(\frac{\delta f}{\delta t}\right)_{coll}=\frac{f_{eq}-f}{\tau} \quad [1.31]$$

where the distribution $f_{eq}$ is the Maxwell–Boltzmann equilibrium distribution (see section 1.5).

In particular, in plasmas, the model generally consists of considering only collisions between plasma particles (mainly electrons) and neutral atoms. In this case, the distribution of electrons gradually tends to the distribution of neutral atoms and we have:

$$\left(\frac{\delta f}{\delta t}\right)_{coll}=\frac{f_n-f}{\tau} \quad [1.32]$$

where $f_n$ is the distribution function of neutral atoms.

This model was proposed by Bhatnagar et al. (1954), and is called the BGK approximation or $\tau$ approximation. It is very useful because of its simplicity. Indeed, under the BGK approximation, the Boltzmann equation can be solved in an approximate way with the same methods as the Vlasov equation.

### 1.3.3. *Boltzmann collision integral*

It is possible to calculate the collisional term of the Boltzmann equation in an integral form by making a number of simplifying assumptions, that is, by considering a sparse medium so as to consider only binary collisions (involving only two particles), and short-range interactions between particles. In addition, it will be necessary to make the *molecular chaos* hypothesis (called *Stosszahlansatz* by Paul Ehrenfest). This hypothesis consists of assuming that the velocities of two colliding particles are a priori uncorrelated and independent of their position.

Thus, determining the collision term of the Boltzmann equation is equivalent to calculating the exchange rate of the number of particles in a volume element $d^3rd^3v$ of phase space, during a time interval $dt$, due to *binary* collisions. This *net* exchange rate can be expressed as the sum of a *gain* (counting particles not initially in this volume element and being there after a time interval $dt$) and a *loss* term (accounting for particles, initially being in the volume element and no longer in it after an interval $dt$). The net exchange rate can thus be written as follows:

$$\frac{\delta f}{\delta t} = \left(\frac{\delta f}{\delta t}\right)^{+} - \left(\frac{\delta f}{\delta t}\right)^{-} \qquad [1.33]$$

where the first term represents the gain and the second term the loss of particles due to collisions (binary).

To evaluate the loss term $(\delta f/\delta t)^{-}$, it is possible to consider an $\alpha$-type particle located, before the collision, in the elementary volume $d^3rd^3v$. After collision with another particle, of type $\beta$, the velocity of the particle changes and is found outside the element of velocity $d^3v$. The particle, of type $\beta$, is in the same element $d^3r$ (but in an element of different velocity $dv_1^3$ close to velocity $v_1$). By noting that the term $f_\beta\left(\vec{r},\vec{v_1},t\right)d^3v_1$ represents the number of $\beta$-type particles per unit volume $d^3r$ (see section 1.2), the flow of the incident beam relating to $\beta$-type particles can be written as follows:

$$\begin{aligned}\Gamma_\beta &= f_\beta\left(\vec{r},\vec{v}_1,t\right)d^3v_1\left|\vec{v_1}-\vec{v}\right| \\ &= f_\beta\left(\vec{r},\vec{v_1},t\right)d^3v_1\cdot\vec{g}\end{aligned} \qquad [1.34]$$

where $\vec{g}\equiv\left|\vec{v_1}-\vec{v}\right|$. Let us assume that the $\beta$-type particle approaches with an impact parameter[6] between $b$ and $b+db$ on a collision plane defined between the angles $\varepsilon$ and $\varepsilon+d\varepsilon$. Multiplying the flow by the surface element $b\cdot db\cdot d\varepsilon$ as well as by the time interval $dt$, the result is none other than the number of particles passing through the surface element $b\cdot db\cdot d\varepsilon$ during a time interval $dt$, that is, (Mahfouf 2016):

$$\Gamma_\beta\cdot b\cdot db\cdot d\varepsilon dt = f_\beta\left(\vec{r},\vec{v_1},t\right)d^3v_1\cdot\vec{g}.b\cdot db\cdot d\varepsilon dt \qquad [1.35]$$

6 The impact parameter $b$ is defined as being the perpendicular distance between the trajectory of the incident particle and the center of the field $U(\mathbf{r})$ created by the target particle.

We obtain the number of $\beta$-type particles which collide with $\alpha$-type particles during the time interval $dt$ by multiplying by the distribution of $\alpha$-type particles. Integrating this quantity over all the possible values of $v_1$, $b$, and $\varepsilon$ of all the $\beta$-type particles, we get the loss rate:

$$\left(\frac{\delta f}{\delta t}\right)_\alpha^- = f_\alpha(\vec{r},\vec{v},t)\sum_\beta \int_{v_1}\int_b\int_\varepsilon f_\beta(\vec{r},\vec{v}_1,t)\,d^3v_1\cdot\vec{g}bdbd\varepsilon \quad [1.36]$$

Following exactly the same reasoning, we obtain the exchange gain, due to collisions, as follows:

$$\left(\frac{\delta f}{\delta t}\right)_\alpha^+ = f_\alpha\left(\vec{r},\vec{v'},t\right)\sum_\beta \int_{v_1'}\int_b\int_\varepsilon f_\beta\left(\vec{r},\vec{v'_1},t\right)d^3v_1'\cdot\vec{g'}\,bdbd\varepsilon \quad [1.37]$$

Assuming that there is a spherical symmetry, we can write:

$$\vec{g'} = \left|\vec{v'_1}-\vec{v'}\right| = |\vec{v_1}-\vec{v}| = \vec{g} \quad [1.38]$$

We also have:

$$d^3v'd^3v_1' = |J|d^3vd^3v_1 \quad [1.39]$$

where, assuming that there is no correlation between velocities or between velocities and positions, the Jacobian is equal to 1. Thus, by combining the loss [1.36] and gain [1.37] terms, we have:

$$\left(\frac{\delta f}{\delta t}\right)_{\text{coll}} = \sum_\beta \int_{v_1'}\int_\Omega \left(f_\alpha' f_\beta' - f_\alpha f_\beta\right)d^3v_1 g\sigma(\Omega,\varepsilon)d\Omega \quad [1.40]$$

where $\sigma(\Omega,\varepsilon)$ is what is called a *differential cross-section* and defined as follows:

$$\sigma(\Omega,\varepsilon)d\Omega = b.db\cdot d\varepsilon \quad [1.41]$$

Now combining the collision term [1.40] with the Boltzmann equation [1.24], we have:

$$\frac{\partial f}{\partial t} + \vec{v}.\nabla f + \vec{a}.\nabla_v f = \sum_\beta \int_{v_1'}\int_\Omega \left(f_\alpha' f_\beta' - f_\alpha f_\beta\right)d^3v_1 g\sigma(\Omega,\varepsilon)d\Omega \quad [1.42]$$

Note that in the case of a single type of particle, the equation becomes:

$$\frac{\partial f}{\partial t}+\vec{v}.\nabla f+\vec{a}.\nabla_v f=\int_{v_1'}\int_{\Omega}\left(f'f_1'-ff_1\right)d^3v_1 g\sigma(\Omega,\varepsilon)d\Omega \qquad [1.43]$$

The right-hand side of the preceding equation is the *Boltzmann collision integral.* Equation [1.43] is an integer-differential equation, that is, involving both derivatives and integrals of the distribution function. Due to its complexity, this equation is not of much practical use and we will not see it in Part 2, which is devoted to applications. However, it has an important theoretical interest, which will be seen in section 1.6 because it makes it possible to understand the relaxation of systems toward the equilibrium state, and makes it possible to demonstrate a fundamental theorem of statistical physics, namely the H-theorem.

### 1.3.4. *Fokker–Planck collision term*

In the case of plasmas, another approximation of the collision term can be of great use. Indeed, when the plasma is strongly ionized, collisions can no longer be considered as binary, because they involve a large number of particles at a time.

A suitable approach in this case is that of *Fokker–Planck,* the aim of which is to describe simultaneous interactions due to Coulomb interactions, assuming, however, small deflection angles.

Consider a quantity $\chi(\vec{v})$ which depends on the relative velocity between two particles of type $\alpha$ and $\beta$. Using the Boltzmann collision integral [1.40], the variation in time of this quantity can then be written as follows:

$$\begin{aligned}\int_v \chi(\vec{v})\left(\frac{\delta f_\alpha}{\delta t}\right)_{coll} d^3v &= \int_{v_1}\int_v\int_\Omega \left(f_\alpha' f_\beta' - f_\alpha f_\beta\right)\chi(\vec{v})g\sigma(\Omega)d\Omega d^3v_1 d^3v \\ &= \int_{v_1}\int_v\int_\Omega f_\alpha f_\beta\left(\chi'-\chi\right)g\sigma(\Omega)d\Omega d^3v_1 d^3v\end{aligned} \qquad [1.44]$$

where $\chi'$ is the function after collision. Assuming there are small deflection angles, we can write:

$$\begin{aligned}\vec{v'} &= \vec{v}+\Delta\vec{v} \\ \chi'\left(\vec{v'}\right) &= \chi(\vec{v}+\Delta\vec{v})\end{aligned} \qquad [1.45]$$

By developing $\chi(\vec{v}+\Delta\vec{v})$ around $\vec{v}$, in series, we have:

$$\chi(\vec{v}+\Delta\vec{v}) = \chi(\vec{v}) + \sum_i \frac{\partial \chi}{\partial v_i}\Delta v_i + \frac{1}{2}\sum_{ij}\frac{\partial^2 \chi}{\partial v_i \partial v_j}\Delta v_i \Delta v_j + \dots \qquad [1.46]$$

By replacing the relationship [1.46] in the relationship [1.44], we then have:

$$\begin{aligned}\int_v \chi(\vec{v})\left(\frac{\delta f_\alpha}{\delta t}\right)_{coll} d^3v = \int_{v_1}\int_v\int_\Omega f_\alpha f_\beta \Bigg(\sum_i \frac{\partial \chi}{\partial v_i}\Delta v_i \\ + \frac{1}{2}\sum_{ij}\frac{\partial^2 \chi}{\partial v_i \partial v_j}\Delta v_i \Delta v_j\Bigg) g\sigma(\Omega) d\Omega d^3v_1 d^3v\end{aligned} \qquad [1.47]$$

By doing an integration by parts, we can write:

$$\begin{aligned}\int_v \chi(\vec{v})\left(\frac{\delta f_\alpha}{\delta t}\right)_{coll} d^3v &= -\int_{v_1}\int_v\int_\Omega \chi \sum_i \frac{\partial \chi}{\partial v_i}\left[\Delta v_i f_\alpha g\sigma(\Omega)d\Omega\right] f_\beta d^3v_1 d^3v \\ &+ \int_{v_1}\int_v\int_\Omega \frac{1}{2}\chi \sum_{ij}\frac{\partial^2 \chi}{\partial v_i \partial v_j}\left[\Delta v_i \Delta v_j f_\alpha g\sigma(\Omega)d\Omega\right] f_\beta d^3v_1 d^3v \\ &= -\int_v \chi \left[\sum_i \frac{\partial \chi}{\partial v_i}\left(f_\alpha \int_{v_1}\int_\Omega \Delta v_i g\sigma(\Omega) d\Omega f_\beta d^3v_1\right)\right] d^3v \\ &+ \int_v \chi \left[\frac{1}{2}\sum_{ij}\frac{\partial^2 \chi}{\partial v_i \partial v_j}\left(f_\alpha \int_{v_1}\int_\Omega \Delta v_i \Delta v_j g\sigma(\Omega) d\Omega f_\beta d^3v_1\right)\right] d^3v\end{aligned} \qquad [1.48]$$

Let us now define the following average quantities:

$$\begin{aligned}&< \Delta v_i >= \int_{v_1}\int_\Omega \Delta v_i g\sigma(\Omega) d\Omega f_\beta d^3v_1 \\ &< \Delta v_i \Delta v_j >= \int_{v_1}\int_\Omega \Delta v_i \Delta v_j g\sigma(\Omega) d\Omega f_\beta d^3v_1\end{aligned} \qquad [1.49]$$

By substituting the mean values [1.49] of the relationship [1.48], we get:

$$\begin{aligned}\int_v \chi \left(\frac{\delta f_\alpha}{\delta t}\right)_{coll} d^3v = -\int_v \chi \Bigg[\sum_i \frac{\partial \chi}{\partial v_i}(f_\alpha < \Delta v_i >) \\ + \frac{1}{2}\sum_{ij}\frac{\partial^2 \chi}{\partial v_i \partial v_j}(f_\alpha < \Delta v_i \Delta v_j >)\Bigg] d^3v\end{aligned} \qquad [1.50]$$

What is interesting about the last relationship is that it is valid for any arbitrary function $\chi(\vec{v})$. In particular, the relationship is valid for $\chi = 1$. In this case, we have:

$$\left(\frac{\delta f_\alpha}{\delta t}\right)_{coll} = -\sum_i \frac{\partial}{\partial v_i}(f_\alpha < \Delta v_i >) + \frac{1}{2}\sum_{ij} \frac{\partial^2}{\partial v_i \partial v_j}(f_\alpha < \Delta v_i \Delta v_j >) \qquad [1.51]$$

This is the *Fokker–Planck* collision.

Quantities $< \Delta v_i >$ and $< \Delta v_i \Delta v_j >$ are called the *coefficient of dynamic friction* and the *diffusion coefficient in the space of velocities*, respectively. Thus, the first term in the relationship [1.51] represents the variation of the mean velocity of a particle, due to Coulomb collisions, while the second term represents the diffusion coefficient of velocities. The Fokker–Planck collision term is of great use in the study of plasmas (Mahfouf 2016) and can be implemented quite easily in numerical simulations.

We can remark something useful before concluding this section: we must not let the technical difficulty (all relative) of the Boltzmann equation (or the Vlasov or Fokker–Planck equation) hide its conceptual simplicity. While it is true that the Boltzmann equation, or even Vlasov equation, are generally not analytically soluble, there are approximate methods, essentially based on the perturbation theory, for solving them. These methods consist of considering small perturbations around the equilibrium values. Such methods make using the Boltzmann equation relatively simple while also revealing the power of the kinetic approach to describe very complex physical scenarios. It is even quite remarkable that a (relatively simple) equation, like the Vlasov equation, describes with astonishing accuracy a large spectrum of physical systems, ranging from plasmas to gravitational systems. We will see this in more detail in Part 2, where applications in different physical systems will be presented.

## 1.4. Steady state: Boltzmann distribution

What will interest us most often in processing the Boltzmann equation is considering perturbations around the equilibrium. It is also necessary to know the distribution function $f$ when the system is at equilibrium. Such a distribution can easily be established, in several different ways; for example, by invoking the *principle of maximum entropy*, by maximizing entropy with the constraints related to the normalization of probabilities and the conservation of total energy (canonical set). The establishment of this distribution can be found in any introductory book to statistical physics (Ngô and Ngô 2008). Here, we present the steps succinctly.

For simplicity, we consider a discrete case. The same reasoning applies in the continuous case, where discrete sums must be replaced by integrals. The problem is to maximize Boltzmann entropy:

$$S = -k_B \sum_i p_i ln p_i \qquad [1.52]$$

where $p_i$ represents the probabilities of the different energy levels that will be denoted as $E_i$. Thus, the entropy [1.52] must be maximized, while respecting certain constraints. In the canonical set, that is, in the case of a contact system with an environment which plays the role of a heat reservoir (thermostat), these constraints are:

$$\sum_i p_i = 1, \quad \sum_i p_i E_i = E \qquad [1.53]$$

where the first constraint expresses the fact that the probabilities must be normalized (the sum of the probabilities must be equal to 1) and the second, the fact that the average energy is constant. This last constraint actually expresses the fact that the total energy is conserved. With the number of particles $N$ being conserved, the total energy is none other than $N$ times the average energy.

A very useful method when it comes to maximizing a function with constraints is that of *Lagrange multipliers*. The function to be maximized is as follows:

$$\bar{S} := S - \alpha\left(\sum_i p_i - 1\right) - \beta\left(\sum_i p_i E_i = E\right) \qquad [1.54]$$

where $\alpha$ and $\beta$ are called Lagrange multipliers. By maximizing the quantity [1.54], that is, $\delta\bar{S}/\delta p_i = 0$, we have:

$$p_i = \frac{\exp(-\beta E_i)}{Z(\beta)} \qquad [1.55]$$

where $\beta = 1/k_B T$ ($T$ being the temperature and $k_B$ the Boltzmann constant) and $Z(\beta)$ is a normalizing constant, called a *partition function*. In the case of *free* particles (i.e. far from any potential), the energy is only given by the kinetic energy $E = mv^2/2$ and the probability [1.56] therefore has the form of a Gaussian function:

$$f(v) \propto \exp(-mv^2/2k_B T) \qquad [1.56]$$

Another way to find the equilibrium function for free particles, called the *Maxwell–Boltzmann*, consists of noting that during a binary collision between two particles, 1 and 2, there are *invariants*, that is, quantities which remain unchanged, before and after collision. Indeed, during the collision of the particle 1 of mass $m_1$ with the particle 2 of mass $m_2$, the mass is conserved:

$$m_1 + m_2 = m_1 + m_2 \qquad [1.57]$$

Similarly, momentum and kinetic energy are also conserved during the collision.

Thus, by denoting the velocities of the particles 1 and 2, before and after collision, as $\vec{v_{1,2}}$ and $\vec{v'_{1,2}}$ the conservation of the quantity of movement is written as:

$$m_1\vec{v}_1 + m_2\vec{v}_2 = m_1\vec{v}_1 + m_2\vec{v}_2 \qquad [1.58]$$

and the conservation of kinetic energy is written as:

$$\frac{1}{2}m_1 v_1^2 + \frac{1}{2}m_2 v_2^2 = \frac{1}{2}m_1 v_1'^2 + \frac{1}{2}m_2 v_2'^2 \qquad [1.59]$$

As a result of this conservation, we have:

$$f_1\left(\vec{v_1}\right) f_2\left(\vec{v_2}\right) = f_1'\left(\vec{v_1}'\right) f_2'\left(\vec{v_2}'\right) \qquad [1.60]$$

or even, by applying the logarithm to the preceding relationship, we have:

$$\ln f_1\left(\vec{v_1}\right) + \ln f_2\left(\vec{v_2}\right) = \ln f_1'\left(\vec{v_1}'\right) + \ln f_2'\left(\vec{v_2}'\right) \qquad [1.61]$$

Thus, by focusing on particle 1, the quantity $\ln f_1$ is said to be invariant of the collision (the same obviously applies to particle 2). This means that this quantity can be written as a linear combination of the invariants of the collision, namely, mass, momentum and kinetic energy.

Thus, we have[7]:

$$\ln f_1\left(\vec{v}_1\right) = m_1\left(a_0 + \vec{a_1}\cdot\vec{v_1} - \frac{1}{2}a_2 v_1^2\right) \qquad [1.62]$$

7 Note that fact $-1/2$ in relationship [1.62] is not necessary. It is chosen here in order to simplify the relationship that follows.

where $a_0$, $\vec{a_1}$ and $a_2$ are constants. Expanding on the relationship [1.62], we obtain:

$$\begin{aligned}\ln f_1(\vec{v}_1) &= m_1\left[a_0+\left(a_{1x}^2+a_{1_y}^2+a_{1_z}^2\right)/(2a_2)\right]\\&-\frac{1}{2}m_1a_2\left[(v_x-a_{1x}/a_2)^2+(v_y-a_{1y}/a_2)^2+(v_z-a_{1z}/a_2)^2\right]\\&=m_1\left[a_0+a_1^2/(2a_2)\right]-\frac{1}{2}m_1a_2\left(\vec{v_1}-\vec{a_1}/a_2\right)^2\end{aligned}\qquad[1.63]$$

where $a_{1x}$, $a_{1y}$ and $a_{1z}$ represent the components of the vector $\vec{a_1}$ along the $x$, $y$ and $z$ axes. At this stage, it may be useful to define the following scalar quantity:

$$\ln C = m_1\left[a_0+a_1^2/(2a_2)\right] \iff C=\exp\left\{m_1\left[a_0+a_1^2/(2a_2)\right]\right\}\qquad[1.64]$$

and the following vector quantity:

$$\vec{v_0}=\frac{\vec{a_1}}{a_2}\qquad[1.65]$$

By replacing these quantities in the relationship [1.63], we get:

$$f(\vec{v_1})=C\exp\left[-\frac{1}{2}m_1a_2\left(\vec{v_1}-\vec{v_0}\right)^2\right]\qquad[1.66]$$

which is none other than the Maxwell–Boltzmann distribution (a Gaussian distribution).

What remains is determining the constants $C$, $a_2$ and $\vec{v_0}$. For this, it should first be noted that the constant $C$ appears in the distribution as a multiplying factor whose role is to normalize the distribution. In accordance with the definition [1.19] given in section 1.2, it would be desirable for the integral of the distribution over the space of velocities to give the density of particles[8], namely:

$$n_1=C\int_{v_1}\exp\left[-\frac{1}{2}m_1a_2\left(\vec{v_1}-\vec{v_0}\right)^2\right]d^3v_1\qquad[1.67]$$

Introducing the following notation:

$$A\equiv m_1a_2/2\quad\text{and}\quad \xi_i\equiv(v_i-v_{1,i}),\quad(i=1,2,3)\qquad[1.68]$$

8 In some cases, it may be useful to normalize the distribution to mass density (mass per unit volume) or even to charge density (charge per unit volume).

The relationship [1.67] is then rewritten as:

$$n_1 = C \iiint_{-\infty}^{+\infty} \exp\left[-A\left(\xi_x^2 + \xi_y^2 + \xi_z^2\right)\right] d\xi_x d\xi_y d\xi_z \qquad [1.69]$$

By calculating the integral (see calculation details in section 1.5), we have:

$$n_1 = C\left(\frac{\pi}{A}\right)^{3/2} = C\left(\frac{2\pi}{m_1 a_2}\right)^{3/2} \qquad [1.70]$$

It should also be noted that the mean velocity is defined as follows:

$$\vec{u} = <\vec{v}> = \frac{1}{n}\int_v f\vec{v}d^3v \qquad [1.71]$$

Using the distribution [1.66], we have:

$$\vec{u} = \frac{C}{n_1}\int_{v_1} \vec{v_1} \exp\left[-\frac{1}{2}m_1 a_2 \left(\vec{v_1} - \vec{v_0}\right)^2\right] d^3 v_1 \qquad [1.72]$$

or, using the notations [1.68]:

$$\vec{u} = \frac{C}{n_1}\iiint_{-\infty}^{+\infty} \left(\xi_x\vec{x} + \xi_y\vec{y} + \xi_z\vec{z}\right) \exp\left[-A\left(\xi_x^2 + \xi_y^2 + \xi_z^2\right)\right] d\xi_x d\xi_y d\xi_z$$
$$+ \frac{C}{n_1}\vec{v_0}\iiint_{-\infty}^{+\infty} \exp\left[-A\left(\xi_x^2 + \xi_y^2 + \xi_z^2\right)\right] d\xi_x d\xi_y d\xi_z \qquad [1.73]$$

The first triple integral is canceled because it corresponds to the integral of a pair function over a symmetrical interval. The second triple integral, on the other hand, is equal to $n_1/C$.

So, we have:

$$\vec{u} = \vec{v_0} \qquad [1.74]$$

Thus, the velocity $\vec{v_0}$ is the mean velocity. It might be useful to note that:

$$\vec{v} = \vec{u} + \vec{c} \qquad [1.75]$$

where $\vec{u}$ is the mean velocity and $\vec{c}$ is the *particle velocity*.

Finally, by defining temperature in a kinetic direction[9], as a measure of average kinetic energy (see also section 1.5), we can write:

$$\begin{aligned}\frac{3}{2}n_1k_BT &= \frac{1}{2}n_1m_1 < c^2 >= \frac{1}{2}m_1\int_{v_1} fc^2d^3v_1\\ &= \frac{1}{2}n_1C\int_c c^2\exp\left(-Ac^2\right)d^3c\end{aligned} \quad [1.76]$$

Or, after integration:

$$\begin{aligned}C &= n_1\left(\frac{\mathrm{m}_1}{2\pi k_BT}\right)^{3/2}\\ a_2 &= \frac{1}{k_BT}\end{aligned} \quad [1.77]$$

Using the relationships [1.70] and [1.77], we can determine the constants as follows:

$$f(c) = n\left(\frac{m}{2\pi k_BT}\right)^{3/2}\exp\left(\frac{-mc^2}{2k_BT}\right) \quad [1.78]$$

This is the Maxwell–Boltzmann distribution, characteristic of systems in thermodynamic equilibrium. However, a system may only have a *local* equilibrium. Thus, the Maxwell–Boltzmann distribution is only valid locally. In these cases, the temperature and mean velocity may vary over time and space. We will then write:

$$f(\vec{r},\vec{v},t) = n(\vec{r},t)\left(\frac{m}{2\pi k_BT(\vec{r},t)}\right)^{3/2}\exp\left(\frac{-m[\vec{v}-\vec{u}(\vec{r},t)]^2}{2k_BT(\vec{r},t)}\right) \quad [1.79]$$

It is important to note that, to obtain the Maxwell–Boltzmann distribution, we considered free particles *libres*, far from any potential. The system may be under the action of an *external* force $\vec{F}(\vec{r})$. If $\vec{F}(\vec{r})$ represents a conservative force, derived from a potential $U(\vec{r})$, either:

$$\vec{F}(\vec{r}) = -\vec{\nabla}U(\vec{r}) \quad [1.80]$$

The density $n(\vec{r})$ will depend on the position through the potential $U(\vec{r})$ and will be given as:

$$n(\vec{r}) = n_0\exp\left(\frac{-U(\vec{r})}{k_BT}\right) \quad [1.81]$$

9 It is very often useful to think of temperature in terms of energy, that is, multiplied by the Boltzmann constant, that is, $k_BT$.

where the term:

$$\exp\left(\frac{-U(\vec{r})}{k_B T}\right) \qquad [1.82]$$

is called the *Boltzmann factor*. The distribution [1.79], in this case, will be written as:

$$f(\vec{r},v) = n_0\left(\frac{m}{2\pi k_B T}\right)^{3/2} \exp\left(\frac{-mc^2}{2k_B T}\right)\exp\left(\frac{-U(\vec{r})}{k_B T}\right) \qquad [1.83]$$

Thus, it is always a question of a Maxwell–Boltzmann distribution of velocities but which is, in this case, multiplied by the Boltzmann factor.

In the case of a (not magnetized) plasma, the force field $\vec{F}(\vec{r})$ typically represents a Coulomb force due to an external electric field $\vec{E}$. This is the result of an electrostatic potential $\Phi_e$, namely:

$$\vec{E} = -\vec{\nabla}\Phi_e(\vec{r}) = -\vec{\nabla}\left\{\frac{U(\vec{r})}{q}\right\} \qquad [1.84]$$

$q$ is the charge of the particle. This field causes the spatial distribution of the particles to no longer be homogeneous. Thus, we have:

$$n(\vec{r}) = n_0 \exp\left[-\frac{q\Phi_e(\vec{r})}{k_B T}\right] \qquad [1.85]$$

Similarly, in the case of a gravitational system, the force field derives from a gravitational potential $\Phi_g$:

$$\vec{g} = -\vec{\nabla}\Phi_g(\vec{r}) = -\vec{\nabla}\left\{\frac{U(\vec{r})}{m}\right\} \qquad [1.86]$$

and in this case, we have:

$$n(\vec{r}) = n_0 \exp\left[-\frac{m\Phi_g(\vec{r})}{k_B T}\right] \qquad [1.87]$$

## 1.5. Maxwell–Boltzmann distribution: properties

In the case of free particles, far from any potential, the equilibrium distribution is therefore that of Maxwell–Boltzmann, which is written as follows:

$$f(v) = n_0 \left(\frac{m}{2\pi k_B T}\right)^{3/2} \exp\left(-mv^2/2k_B T\right)$$
$$= n_0 \left(\frac{m}{2\pi k_B T}\right)^{3/2} \exp\left(-v^2/v_{th}^2\right) \qquad [1.88]$$

where $v_{th} := \sqrt{2k_B T/m}$ is called the *thermal velocity*. Also, it is important to know the properties of this distribution:

– Let us firstly observe that the distribution [1.88] does not depend on the velocity vector but only on $v^2$, that is, on its modulus. Thus, if the distribution were expressed in spherical coordinates, it would thus contain no information on the angles $\theta$ and $\phi$ defining the direction of the velocity vector. This means that it is the same, whatever the direction; it is said to be *isotropic*.

– It can also be noted that the distribution [1.88] in 3D is *separable*, that is, reduced to a product of one-dimensional (1D) distributions. Indeed, we have:

$$f(v) = f(v_x) f(v_y) f(v_z) \qquad [1.89]$$

where the 1D distribution is given by:

$$f(v_x) = \left(\frac{m}{2\pi k_B T}\right)^{1/2} \exp\left(-mv_x^2/2k_B T\right) = \left(\frac{m}{2\pi k_B T}\right)^{1/2} \exp\left(-v_x^2/v_{th}^2\right) \qquad [1.90]$$

Property [1.89] expresses the independence of the different velocity components. Thus, the probability that a particle has a given velocity along the $x$-axis, for example, is independent of the probability that it has a given velocity along the $y$- or $z$-axis.

Also, using the Gaussian integral[10]:

$$\int_{-\infty}^{\infty} \exp\left(-x^2\right) dx = \sqrt{\pi} \qquad [1.94]$$

10 In general, we have:

$$\int_{-\infty}^{\infty} e^{-\alpha(x-b)^2} dx = \sqrt{\frac{\pi}{\alpha}} \quad \int_{-\infty}^{\infty} e^{-(x-\langle x\rangle)^2/2\sigma^2} dx = \sqrt{2\pi\sigma^2} \qquad [1.91]$$

The integrals of the following type are also given:

$$I_m(\alpha) = \int_0^{+\infty} x^m e^{-ax^2} dx \qquad [1.92]$$

it can easily be shown that it is normalized[11], that is:

$$\int_{-\infty}^{\infty}\int_{-\infty}^{\infty}\int_{-\infty}^{\infty} f(v)dv_x dv_y dv_z = n_0 \qquad [1.97]$$

It is also important to know the *moments*, that is the mean values of the whole power of the velocity, that is, $\langle v^n \rangle$. The latter represent statistical indicators that provide information on distribution. In particular, two moments are important, namely the moment of order 1 ($\langle v \rangle$), that is the mean velocity, and the moment of order 2 ($\langle v^2 \rangle$), which gives information on the mean square error.

The mean velocity (in modulus) is defined, for a 3D distribution, as follows:

$$\langle v \rangle = \frac{1}{n_0}\int_{-\infty}^{\infty}\int_{-\infty}^{\infty}\int_{-\infty}^{\infty} v f(\mathbf{v})dv_x dv_y dv_z \qquad [1.98]$$

Since the Maxwell–Boltzmann distribution is isotropic, it is useful to use spherical coordinates $(v, \theta, \phi)$ instead of Cartesian coordinates $(v_x, v_y, v_z)$, and write:

$$\begin{aligned}\langle v \rangle &= (m/2\pi k_B T)^{3/2}\int_0^{\infty}\int_0^{\pi}\int_0^{2\pi} v\exp\left(-v^2/v_{th}^2\right) v^2 dv sin(\theta)d\theta d\phi \\ &= (m/2\pi k_B T)^{3/2}\int_0^{\infty} v\exp\left(-v^2/v_{th}^2\right) 4\pi v^2 dv \\ &= \left(\pi v_{th}^2\right)^{-3/2} 4\pi v_{th}^4 \int_0^{\infty}\left[\exp\left(-y^2\right)\right] y^3 dy\end{aligned} \qquad [1.99]$$

Observing that:

$$\int_0^{\infty}\left[\exp\left(-y^2\right)\right] y^3 dy = \frac{1}{2} \qquad [1.100]$$

---

By differentiating with respect to $\alpha$ and integrating by parts, the following recurrence relation is obtained $I_{m+2}(\alpha) = \frac{m+1}{2\alpha} I_m(\alpha)$. We get the first terms by:

$$\begin{aligned}&I_0(\alpha) = \frac{1}{2}\sqrt{\frac{\pi}{\alpha}}, \quad I_2(\alpha) = \frac{1}{4\alpha}\sqrt{\frac{\pi}{\alpha}}, \quad I_4(\alpha) = \frac{3}{8\alpha^2}\sqrt{\frac{\pi}{\alpha}} \\ &I_1(\alpha) = \frac{1}{2\alpha}, \quad I_3(\alpha) = \frac{1}{2\alpha^2}, \quad I_5(\alpha) = \frac{1}{\alpha^3}\end{aligned} \qquad [1.93]$$

11 It should be noted that it is possible, in an equivalent manner, to define a distribution $\hat{f}(v) = f(v)/n_0$. This would be normalized to the unit, that is,

$$\int_{-\infty}^{\infty}\int_{-\infty}^{\infty}\int_{-\infty}^{\infty} \hat{f}(v)dv_x dv_y dv_z = 1 \qquad [1.95]$$

In this case, the average values would be:

$$\langle v^n \rangle = \int_{-\infty}^{\infty}\int_{-\infty}^{\infty}\int_{-\infty}^{\infty} v^n \hat{f}(v)dv_x dv_y dv_z \qquad [1.96]$$

the mean velocity can be expressed as follows:

$$\langle v\rangle = 2v_{th}/\sqrt{\pi} = \sqrt{\frac{8k_BT}{\pi m}} \qquad [1.101]$$

We can also look at the mean value of the velocity on a given axis, say, the $x$-axis. In order to do this, it is important to distinguish between the mean norm of the velocity in a given direction and the mean component in this direction.

Thus, if we are interested in the projection of the velocity vector on a given axis, we will write:

$$\begin{aligned}\langle v_x\rangle =& \frac{1}{n_0}\int |v_x| f(\mathbf{v})d^3v \\ =& \left(\frac{m}{2\pi k_BT}\right)^{3/2}\int_{-\infty}^{\infty} dv_y \exp\left(\frac{-v_y^2}{v_{th}^2}\right)\int_{-\infty}^{\infty} dv_z \exp\left(\frac{-v_z^2}{v_{th}^2}\right) \\ &\times \int_{-\infty}^{\infty} v_x \exp\left(\frac{-v_x^2}{v_{th}^2}\right) dv_x = 0\end{aligned} \qquad [1.102]$$

because the last integral represents the integral of an odd function over an asymmetric interval. Of course, the same is true for the other two components, $\langle v_x\rangle = \langle v_y\rangle = \langle v_z\rangle = 0$. This is a consequence of the isotropy of the distribution. We can understand this by *reductio ad absurdum*; let us suppose that the mean value along an axis, say $\langle v_x\rangle$, is not zero. This would mean that a particle would have a greater chance of going in one direction of the $x$-axis than in the other direction, which would be in contradiction with the isotropy of the distribution.

If, on the other hand, we are interested in the absolute value of this quantity, we have:

$$\begin{aligned}\langle v_x\rangle =& \frac{1}{n_0}\int |v_x| f(\mathbf{v})d^3v \\ =& \left(\frac{m}{2\pi k_BT}\right)^{3/2}\int_{-\infty}^{\infty} dv_y \exp\left(\frac{-v_y^2}{v_{th}^2}\right)\int_{-\infty}^{\infty} dv_z \exp\left(\frac{-v_z^2}{v_{th}^2}\right) \\ &\times \int_{0}^{\infty} 2v_x \exp\left(\frac{-v_x^2}{v_{th}^2}\right) dv_x\end{aligned} \qquad [1.103]$$

By observing that the first two integrals are each equal to $\sqrt{\pi}v_{th}$ and that the last integral is equal to $v_{th}^2$, we obtain the mean velocity norm in a given direction as follows:

$$\begin{aligned}\langle |v_x| \rangle &= \left(\pi v_{th}^2\right)^{-3/2} \pi v_{th}^4 \\ &= (\pi)^{-1/2} v_{th} = \left(\frac{2k_B T}{\pi m}\right)^{1/2}\end{aligned} \qquad [1.104]$$

Another quantity, just as important, is the moment of order 2. This quantity is doubly interesting; statistically speaking, the moment of order 2, or more precisely the mean square error, gives information on the width of the distribution or, in other words, on the uncertainty on the random variable in question (here, the velocity). Physically speaking, this quantity is also important because, apart from a multiplicative factor, it defines the mean kinetic energy. In 1D, we have:

$$\begin{aligned}\langle v^2 \rangle &= \frac{1}{n_0} \int_{-\infty}^{\infty} \left(\frac{m}{2\pi k_B T}\right)^{1/2} v^2 \exp\left(-\frac{v^2}{v_{th}^2}\right) dv \\ &= \int_{-\infty}^{\infty} \left(\frac{m}{2\pi k_B T}\right)^{1/2} v_{th}^3 y^2 \exp\left(-y^2\right) dy \\ &= \left(\frac{m}{2\pi k_B T}\right)^{1/2} v_{th}^3 \int_{-\infty}^{\infty} y^2 \exp\left(-y^2\right) dy\end{aligned} \qquad [1.105]$$

By doing an integration by parts, we have:

$$\int_{-\infty}^{\infty} y^2 \exp\left(-y^2\right) dy = \left[-\frac{1}{2} y \exp\left(-y^2\right)\right]_{-\infty}^{\infty} - \int_{-\infty}^{\infty} -\frac{1}{2} \exp\left(-y^2\right) dy = \frac{\sqrt{\pi}}{2} \qquad [1.106]$$

Then, equation [1.105] leads to:

$$\langle v^2 \rangle = \frac{v_{th}^2}{2} = \frac{k_B T}{m} \qquad [1.107]$$

In 3D, we can easily notice that integral [1.105] repeats thrice and we simply have:

$$\langle v^2 \rangle = \frac{3 v_{th}^2}{2} = 3 \frac{k_B T}{m} \qquad [1.108]$$

By multiplying the two previous equations by the mass and dividing by two, we obtain the average value of the kinetic energy $\langle E \rangle \equiv m \langle v^2 \rangle / 2$, that is:

$$\langle E \rangle = 3 k_B T \qquad [1.109]$$

Comparing with the relationship [1.107], it is easy to see that the mean total value of the kinetic energy (on the three axes) is three times the mean energy along only one axis. This is indeed where the factor 3 in the mean energy value [1.110] comes from. This is called the *equipartition theorem of kinetic energy*. Kinetic energy is "stored" equitably between the three axes (this is again a consequence of the isotropy of the Maxwell–Boltzmann distribution).

Note that in general, the mean energy for a system of free particles, at equilibrium, is given by:

$$\langle E \rangle = dk_B T \qquad [1.110]$$

where $d$ is the number of degrees of freedom. Instead of having point particles, if we had molecules with a certain structure, dihydrogen $H_2$ molecules, for example, we would have to take into account the degrees of freedom expressing the transverse displacement of the center of mass as well as the degrees of freedom of rotation, with possible axes of rotation and we would thus have $d = (3\text{transverse displacement}) + 2(\text{rotation}) = 5$.

Table 1.1 summarizes the average quantities that we have just calculated. Also, Figure 1.1 shows examples of the 1D Maxwell–Boltzmann distribution, for different values of thermal velocity (or, equivalently, of temperature). It can be observed that the higher the temperature, the wider the distribution (the mean square error increases with temperature). This reflects the fact that, when the temperature rises, there is more energy available for the particles that make up the system. In particular, the high energies (corresponding in this case to high velocities) become accessible to the particles. Figure 1.2 shows, for its part, examples of the same distribution but with 2D in order to assess the isotropy of the distribution.

| | |
|---|---|
| **Mean of standard** | $\langle \lvert v \rvert \rangle = 2\left(\frac{2k_BT}{\pi m}\right)^{1/2}$ |
| **Mean of the standard on an axis** | $\langle \lvert v_{x,y,z} \rvert \rangle = \left(\frac{2k_BT}{\pi m}\right)^{1/2}$ |
| **Mean on an axis** | $\langle v_x \rangle = 0$ |
| **Mean square error** | $\sqrt{\langle v^2 \rangle} = \left(\frac{3k_BT}{m}\right)^{1/2}$ |

**Table 1.1.** *Table summarizing the mean values of the Maxwell–Boltzmann distribution*

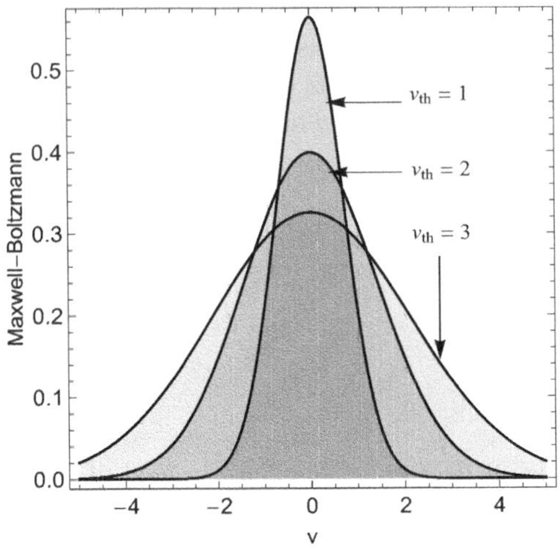


**Figure 1.1.** *Examples of Maxwell–Boltzmann distributions (1D), for different values of the thermal velocity* $v_{th} := \sqrt{2k_BT/m}$ *(arbitrary units). As the thermal velocity (or temperature) increases, the distribution is spread further. For a color version of this figure, see www.iste.co.uk/ourabah/collective.zip*

The Maxwell–Boltzmann distribution is characteristic of systems at equilibrium and is omnipresent in a large number of physical situations, as much in gases or plasmas as in astrophysics for example (velocity distribution of galaxies). Its origin can be understood in a purely statistical way because the Maxwell–Boltzmann distribution is none other than a Gaussian distribution. Now, the ubiquity of the Gaussian distribution can be understood through the *central limit theorem,* which states that a sum of a large number of random variables tends toward a Gaussian. Thus, the size distribution in a given population or that of the intelligence quotient, for example, is close to a Gaussian distribution and has this bell shape characteristic of the Gaussian (see Figure 1.1). This is due to the large number of random variables that contribute to the phenomenon, for example, environmental or genetic factors in the case of size. In physics, random variables are exchanges of energy and momentum between particles that occur at each collision.

To show the validity of the Maxwell–Boltzmann distribution in describing concrete physical situations, Figure 1.3 shows a simulation of the velocity distribution of a 1D solid (Mohazzabi and Shankar 2018). It can be seen that the 1D Maxwell–Boltzmann distribution [1.90] correctly reproduces the observed behavior. Similarly, Figure 1.3 shows the data from a simulation in the case of a 3D solid.

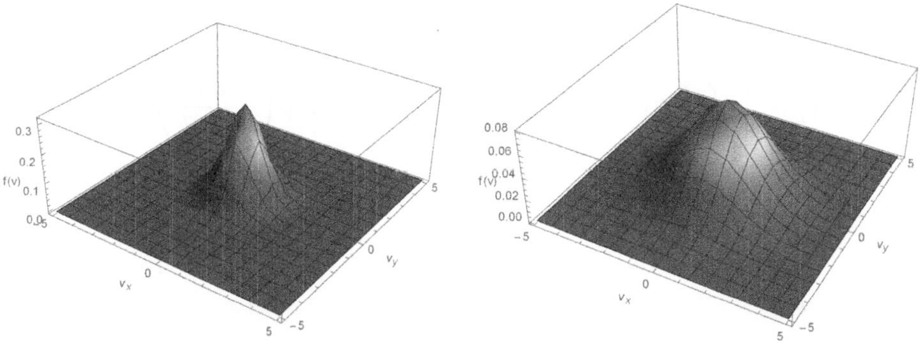


**Figure 1.2.** *Examples of 2D Maxwell–Boltzmann distributions for two different values of the thermal velocity* $v_{th} := \sqrt{2k_B T/m}$. *(a)* $v_{th} = 1$ *and (b)* $v_{th} = 2$ *(arbitrary units). For a color version of this figure, see www.iste.co.uk/ourabah/collective.zip*

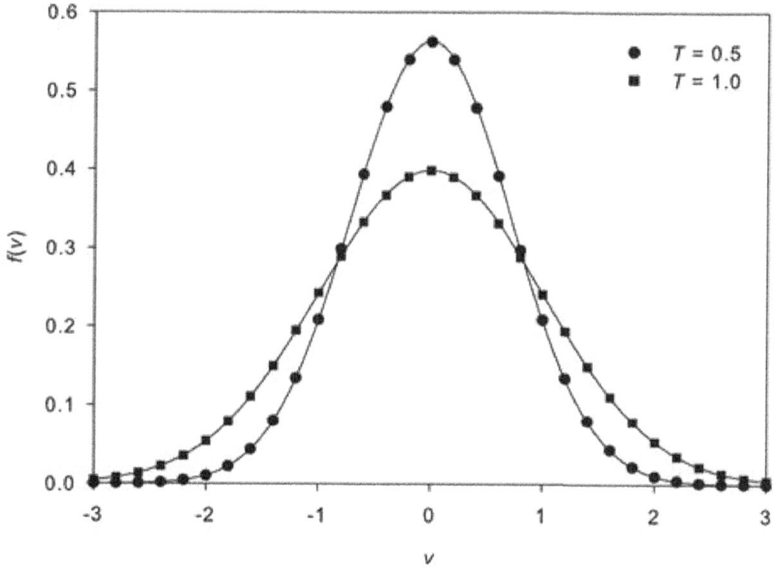


**Figure 1.3.** *Distribution of the velocities of the atoms of a 1D solid, interacting through harmonic potential. Points represent the results of a numerical simulation and the solid curve represents the 1D Maxwell–Boltzmann distribution (source: Mohazzabi and Shankar (2018))*

Again, the velocity distribution (in modulus) is reproduced by the Maxwell–Boltzmann distribution (with the notations used here):

$$P(v) = \frac{1}{n_0}\int\int f(v)d\theta d\phi = 4\pi\left(\frac{m}{2\pi k_B T}\right)^{3/2} v^2 \exp\left(-\frac{mv^2}{2k_B T}\right) \qquad [1.111]$$

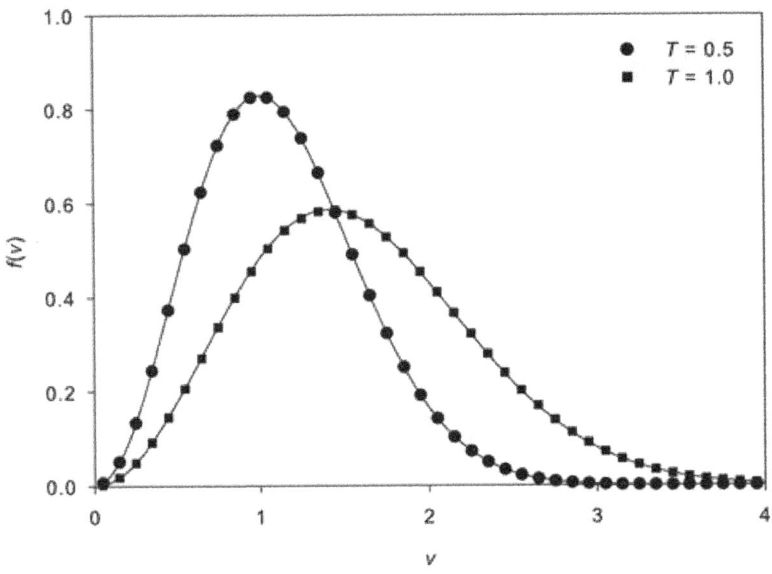


**Figure 1.4.** *The velocity distribution of atoms in a 3D solid, interacting via harmonic potential. Points represent the results of a numerical simulation and the solid curve represents the Maxwell–Boltzmann distribution [1.111] (source: Mohazzabi and Shankar (2018))*

## 1.6. H-theorem

We established the equilibrium distribution in section 1.5. It should be noted that time was not involved in this approach because the state of equilibrium is a steady state.

What remains for us to understand is precisely how time intervenes, that is, how the system evolves from an out-of-equilibrium state to a state of equilibrium. This is called *relaxation*. The Boltzmann equation makes it possible to study this.

To do this, consider a distribution function $f(\vec{v},t)$ (independent of the position) satisfying the Boltzmann equation and, from this distribution, let us build the following quantity:

$$H(t) = \int_v f(\vec{v},t)\ln[f(\vec{v},t)]d^3v \tag{1.112}$$

Quantity [1.112] is intimately related to the Boltzmann entropy [1.52] which, in the continuous case, takes the following form:

$$S = -k_B \int_v \int_r f(\vec{v},t)\ln[f(\vec{v},t)]d^3v d^3r \tag{1.113}$$

Since the distribution $f(\vec{v},t)$ independent of the position $\vec{r}$, the integral on the position in the preceding equation gives the volume of the system $V$. Thus, it can be shown that the quantity $H(t)$ is linked to the entropy by a simple proportionality relation, namely:

$$\frac{S}{V} = -k_B H \tag{1.114}$$

As we are interested in the temporal evolution of the system, let us calculate the derivative of $H(t)$ with respect to time. We have:

$$\frac{\partial H(t)}{\partial t} = \int_v (1+\ln f(\vec{v},t))\frac{\partial f(\vec{v},t)}{\partial t}d^3v \tag{1.115}$$

By replacing the Boltzmann equation [1.43] in the preceding relationship, we have:

$$\frac{\partial H(t)}{\partial t} = \int_\Omega \int_v \int_{v_1} (1+\ln f)\left(f'f_1' - ff_1\right)g\sigma(\Omega)d\Omega d^3v_1 d^3v \tag{1.116}$$

Since the function $H(t)$ can be defined in the same way for $f_1 = f_1(\vec{v_1},t)$, quantities $\vec{v}$ and $\vec{v_1}$ are interchangeable. Thus, we have:

$$\frac{\partial H(t)}{\partial t} = \int_\Omega \int_{v_1} \int_v (1+\ln f_1)\left(f_1'f' - f_1 f\right)g\sigma(\Omega)d\Omega d^3v d^3v_1 \tag{1.117}$$

By adding the two previous equations and dividing by 2, we get:

$$\frac{\partial H(t)}{\partial t} = \frac{1}{2}\int_\Omega \int_v \int_{v_1} (2+\ln ff_1)\left(f_1'f' - f_1 f\right)g\sigma(\Omega)d\Omega d^3v_1 d^3v \tag{1.118}$$

In the same way, it is possible to replace $\vec{v}$ and $\vec{v_1}$ (velocities before the collision) by $\vec{v'}$ and $\vec{v'_1}$ (velocities after the collision). So, we have:

$$\frac{\partial H(t)}{\partial t} = \frac{1}{2} \int_{\Omega} \int_{v} \int_{v_1} (2 + \ln f' f'_1) (f_1 f - f'_1 f') g\sigma(\Omega) d\Omega d^3 v_1 d^3 v \qquad [1.119]$$

We can now combine the two previous relationships and write:

$$\frac{\partial H(t)}{\partial t} = \frac{1}{4} \int_{\Omega} \int_{v} \int_{v_1} \left[ \ln \left( \frac{f f_1}{f' f'_1} \right) \right] (f'_1 f' - f_1 f) g\sigma(\Omega) d\Omega d^3 v_1 d^3 v \qquad [1.120]$$

From this relationship, it is possible to analyze the sign of $\partial H / \partial t$. The following three scenarios can be observed:

1) if $f' f'_1 > f f_1 \Longrightarrow \ln \left( \frac{f f1}{f' f'_1} \right) > 0 \Longrightarrow \frac{\partial H(t)}{\partial t} < 0$. Indeed, this is true because all the other factors of the relationship [1.120] are positive;

2) similarly, if $f' f'_1 < f f_1 \Longrightarrow \ln \left( \frac{f f1}{f' f'_1} \right) > 0 \Longrightarrow \frac{\partial H(t)}{\partial t} < 0$;

3) also, if $f' f'_1 = f f_1 \Longrightarrow \ln \left( \frac{f f1}{f' f'_1} \right) \rightarrow 0 \Longrightarrow \frac{\partial H(t)}{\partial t} = 0$.

In this case, the function $H(t)$ (and, similarly, entropy $S$) is indeed the state of equilibrium.

It can be noted that, in the three possible situations, we always have:

$$\frac{\partial H(t)}{\partial t} \leq 0 \qquad [1.121]$$

This is Boltzmann's *H-theorem*. We learn from this theorem that quantity $H(t)$, defined by relationship [1.112], cannot but increase or remain constant as time passes. Since function $H(t)$ is proportional to entropy $S$ and the proportionality factor is negative (see relationship [1.114]), the theorem tells us that the entropy of system $S$ can only increase or remain constant over time.

Thus, function $H(t)$ decreases with time until it reaches a particular value, characterized by (in which case $\partial H(t) / \partial t = 0$):

$$f' f'_1 = f f_1 \qquad [1.122]$$

Again, of course, we can think in terms of entropy. This value increases over time until it reaches a value characterized by the condition [1.122]. This is the thermodynamic equilibrium condition!

Taking the logarithm of the relationship [1.122], we have:

$$\ln f + \ln f_1 = \ln f' + \ln f_1' \qquad [1.123]$$

which is none other than the relationship [1.61], from which we found the Maxwell–Boltzmann distribution in section 1.4. Thus, the only function satisfying the thermodynamic equilibrium condition:

$$\frac{\partial H(t)}{\partial t} = \frac{\partial S(t)}{\partial t} = 0 \qquad [1.124]$$

is the Maxwell–Boltzmann distribution function.

This is a fundamental theorem of kinetic theory. However, at the time it was formulated by Boltzmann in 1872, strong objections were raised against this theorem, leading to a debate that remained famous (Rothstein 1974; Steckline 1983). This is the historical point that will be discussed in section 1.7.

## 1.7. Paradoxes related to H-theorem

After Boltzmann's publication of the H-theorem, two paradoxes were observed, first by the Austrian physicist *Johann Josef Loschmidt* in 1876, and then by the German mathematician *Ernst Zermelo* in 1896. Here, we will present these paradoxes and how they were resolved.

### 1.7.1. *Loschmidt's paradox of irreversibility*

Loschmidt's paradox started from a rather simple observation: the laws of mechanics, governing the evolution of each particle in the medium, are *reversible*[12]. This seems to contradict the H-theorem according to which there is *irreversibility* (function $H$ can only decrease in time or, in an equivalent way, entropy $S$ can only increase in time).

Thus, Loschmidt observed that if function $H(t)$ increased over time, it would be enough to *exactly* reverse the velocities of each particle in the medium so that function $H(t)$ begins to decrease.

12 This can be seen in Newton's equations: by changing the time variable $t$ to $-t$, the fundamental equations of mechanics remain invariant.

Today, we know, with the advent of chaos theory, that chaotic systems are extremely dependent on initial conditions. Thus, if the particle velocities were reversed, even a small deviation would result in a deviation from the initial orbit reversed exactly. This can be studied via numerical simulations where it is shown that an approximate inversion of velocities does lead to a decrease in the function $H(t)$, as predicted by Loschmidt. However, function $H(t)$ very quickly starts decreasing again. Thus, since the timescale at which the system evolves is much larger than the timescale at which collisions occur, function $H(t)$ decreases overall over time, even if it can fluctuate at smaller timescales.

### 1.7.2. *Zermelo paradox*

Another criticism was made by Zermelo in 1896. He found a contradiction between Boltzmann's H-theorem and the *recurrence theorem*, established by Poincaré in 1890. Indeed, Poincaré's recurrence theorem says that, for all of the initial conditions, a conservative dynamic system whose phase space is of finite volume will repeatedly pass as close as we want to its initial condition. Zermelo then pointed out that this theorem was in contradiction with the existence of a quantity that evolves monotonically over time, as predicted by the H-theorem.

The Poincaré recurrence theorem is indeed valid and of great mathematical rigor. However, it says nothing about the time it takes for the system to pass through the initial conditions again (this is called the recurrence time). Boltzmann's response consists of estimating this recurrence time for a macroscopic system, composed of a very large number ($N$) of particles. Boltzmann estimated this time to be of the order of $10^N$. Thus, considering the typical case of $N \sim 10^{23}$, the Poincaré recurrence time greatly exceeds the age of the Universe. For macroscopic systems, where kinetic theory makes sense, the recurrence theorem can be ignored.

**Figure 1.5.** *From left to right, Ludwig Boltzmann, Johann Josef Loschmidt and Ernst Zermelo*

# 2

# Hydrodynamic Approach

*Although to penetrate into the intimate mysteries of nature and thence to learn the true causes of phenomena is not allowed to us, nevertheless it can happen that a certain fictive hypothesis may suffice for explaining many phenomena.*

Leonhard Euler

In Chapter 1, we established the Boltzmann equation, which is at the heart of kinetic theory. Such a kinetic description is indeed of great precision in the sense that it introduces the notion of distribution and, in doing so, makes it possible to describe the differences (of velocity or energy) between the different particles that constitute the system. We can say that it is a *mesoscopic* description, that is, intermediate between a *microscopic* description (i.e. based on the individual equations of motion of each particle) and a *macroscopic* description. This chapter presents, precisely, such a macroscopic approach, called *hydrodynamic* or *fluid*.

Before formulating the hydrodynamic model, it may be interesting to pause briefly on an important point. Recall that a distribution, for example, the Maxwell–Boltzmann (Gaussian) distribution, is generally characterized by its moments. This means that, if we knew *all the moments* of a given distribution, we could, in principle, reconstruct it. However, many distributions, including the Gaussian, are characterized by an *infinite* number of moments. A macroscopic approach is an approach involving only average values, that is, only the moments of the distribution. Thus, as a distribution is generally characterized by an infinite number of moments, it would in principle be necessary to write an infinity of equations, governing the variation of these moments, to have a description as precise as the kinetic theory, that is, the Boltzmann or Vlasov equation. As this is impossible, the idea is to be content with the evolution of the first moments of the distribution; for example, the moment of order 0 (i.e. the density) and the moment of order 1 (the mean velocity). This is the main idea of the hydrodynamic or fluid approach.

## 2.1. Fluid model: a heuristic approach

Before constructing fluid model equations from kinetic theory, it may be interesting to establish these equations, under certain conditions, in a simple and intuitive way. To do this, consider a "perfect gas" type fluid, that is, where interactions between particles that compose it can be ignored. However, the fluid can be subjected to an external force field $\vec{F}$. We will also assume that there are no collisions between the particles.

Consider a macroscopic volume of this fluid and let $n$ be the particle density (number of particles per unit volume) in this volume. It can then be observed that a variation of the density $n$ is necessarily due to a *flow* of particles crossing the surface which surrounds this volume. We can then write:

$$\frac{\partial n}{\partial t} + \vec{\nabla} \cdot (n\vec{u}) = 0 \qquad [2.1]$$

$\vec{u}$ being the fluid velocity. This equation simply expresses the notion of particle number conservation, that is, the fact that a change in density is caused by a flow of particles passing through the surface surrounding the volume.

Equation [2.1] describes the variation of density $n$ in the medium, but involves the fluid velocity $\vec{u}$. In the presence of a force field, this velocity is not constant but varies under the effect of the force field. To describe this evolution, we can simply apply the fundamental relationship of dynamics to what is called a *fluid particle*, that is, a small volume of the fluid, considered small enough not to deform too much over time, but large enough to contain a macroscopic number of particles. Thus, it will be possible to write:

$$\frac{d\vec{u}}{dt} = \frac{\vec{F}}{m} \qquad [2.2]$$

It should however be noted that the fluid velocity $\vec{u}$ may not be homogeneous in the medium. Thus, the total derivative of $\vec{u}$ can be expanded to explicitly include the variation of fluid velocity along the $x$, $y$ and $z$ directions as follows:

$$\frac{d\vec{u}}{dt} = \frac{\partial \vec{u}}{\partial t} + \frac{\partial x}{\partial t}\frac{\partial \vec{u}}{\partial x} + \frac{\partial y}{\partial t}\frac{\partial \vec{u}}{\partial y} + \frac{\partial z}{\partial t}\frac{\partial \vec{u}}{\partial z} = \frac{\partial \vec{u}}{\partial t} + \vec{u} \cdot \nabla \vec{u} \qquad [2.3]$$

Equation [2.2] can then be written as:

$$\rho\left(\frac{\partial \vec{u}}{\partial t}+\vec{u}\nabla\vec{u}\right)=\vec{\mathscr{F}} \qquad [2.4]$$

where $\rho = mn$ is the mass density and $\vec{\mathscr{F}} = n\vec{F}$ represents a force density.

Equations [2.1] and [2.4] are a simplistic version of the hydrodynamic model. In particular, one essential ingredient is missing from equation [2.4], namely the pressure force that will be discussed in detail in sections 2.4 and 2.5. The main idea of the fluid model, however, remains the same, namely an equation for the conservation of the number of particles and an equation describing the variation of the fluid velocity (or of the quantity of movement). These equations can be established, in a more complete form, from kinetic theory. This will be discussed in sections 2.2 and 2.3.

## 2.2. Macroscopic transport equation

To construct the hydrodynamic or fluid model from kinetic theory, let us recall that a distribution function $f(\vec{\mathrm{r}},\vec{\mathrm{v}},t)$ is a function of space coordinates, velocities and time. Now, the macroscopic magnitudes that we are interested in, for example, density $n$, temperature $T$ and mean velocity $\vec{u}$ are not functions of velocity, but only of space and time; density, for example, may vary at one point over time and may also vary from one point to another, but is not a function of velocity. Intuitively, we would be tempted to integrate the distribution $f(\vec{\mathrm{r}},\vec{\mathrm{v}},t)$ over the velocity space; quantities thus obtained no longer depend on velocities and represent macroscopic magnitudes. Equations describing the variation of these quantities, in space and time, constitute the hydrodynamic model.

For simplicity, we will assume throughout the chapter that collisions are non-existent or that they can be ignored. Thus, our starting point will be the Boltzmann equation without collisions, that is, the Vlasov equation [1.25].

Consider any microscopic quantity as a function of velocity, say $\chi(\vec{\mathrm{v}})$. Since the mean value of this function is obtained by multiplying it by the distribution function $f(\vec{\mathrm{x}},\vec{\mathrm{v}},t)$, and by integrating over the velocity space, the equation describing the variations of this mean value, in space and time, can be obtained in the same way by multiplying the Vlasov equation [1.25] by $\chi(\vec{\mathrm{v}})$ and integrating over the entire velocity space. Thus, we have:

$$\int_v \chi\frac{\partial f}{\partial t}d^3\vec{\mathrm{v}}+\int_v \chi\vec{\mathrm{v}}\cdot\vec{\nabla}_r f d^3\vec{\mathrm{v}}+\int_v \chi\vec{a}\cdot\vec{\nabla}_v f d^3\vec{\mathrm{v}}=0 \qquad [2.5]$$

Let us proceed term by term. Note first that the first term of equation [2.5] can be written as follows:

$$\int_v \chi \frac{\partial f}{\partial t} d^3\vec{v} = \frac{\partial}{\partial t}\left(\int_v \chi f d^3\vec{v}\right) - \int_v f \frac{\partial \chi}{\partial t} d^3\vec{v} \qquad [2.6]$$

It can then be noted that the second term of the right-hand member of this equation is canceled out because $\chi$ is assumed to be a function of velocity only and not of time. Using the definition of the mean value (see, for example, [1.98]), we have:

$$\int_v \chi \frac{\partial f}{\partial t} d^3\vec{v} = \frac{\partial}{\partial t}(n\langle\chi\rangle) \qquad [2.7]$$

Similarly, the second term of equation [2.5] can be written as follows:

$$\begin{aligned}\int_v \chi \vec{v} \cdot \vec{\nabla}_x f d^3\vec{v} &= \vec{\nabla}_x \cdot \int_v \vec{v} \chi f d^3\vec{v} \\ &- \int_v \left(\vec{v} \cdot \vec{\nabla}_x \chi\right) f d^3\vec{v} - \int_v \chi f (\vec{\nabla}_x \cdot \vec{v}) d^3\vec{v}\end{aligned} \qquad [2.8]$$

Terms $\vec{\nabla}_x \cdot \vec{v}$ and $\vec{\nabla}_x \cdot \chi$ cancel each other out, since neither $\chi$ nor $\vec{v}$ are functions of $\vec{x}$ ($\vec{x}$, $\vec{v}$, and $t$ are *independent variables*). We then have:

$$\int_v \chi \vec{v} \cdot \vec{\nabla}_x f d^3\vec{v} = \vec{\nabla}_x \cdot \int_v \vec{v} \chi f d^3\vec{v} = \vec{\nabla}_x \cdot (n\langle\chi \vec{v}\rangle) \qquad [2.9]$$

Similarly, it is possible to write the third term of equation [2.5] as follows:

$$\begin{aligned}\int_v \chi \vec{\gamma} \cdot \vec{\nabla}_v f d^3\vec{v} &= \int_v \vec{\nabla}_v \cdot (\chi \vec{\gamma} f) d^3\vec{v} \\ &- \int_v f (\vec{\gamma} \cdot \vec{\nabla}_v \chi) d^3\vec{v} - \int_v \chi f (\vec{\nabla}_v \cdot \vec{\gamma}) d^3\vec{v}\end{aligned} \qquad [2.10]$$

Note that the last integral vanishes if the component of force along a given direction, say $i$, does not depend on the component of the velocities $v_i$ along this same direction (this is indeed the case for the Lorentz force or the gravitational force).

Note also that the first integral of the right-hand side is a sum of triple integrals in the form:

$$\begin{aligned}\int_v \vec{\nabla}_v \cdot (\chi \vec{\gamma} f) d^3\vec{v} &= \iiint_{-\infty}^{+\infty} \frac{\partial}{\partial v_x}(\gamma_x \chi f) dv_x dv_y dv_z \\ &+ \iiint_{-\infty}^{+\infty} \frac{\partial}{\partial v_y}(\gamma_y \chi f) dv_x dv_y dv_z + \cdots\end{aligned} \qquad [2.11]$$

For each of these integrals, we have:

$$\iiint_{-\infty}^{+\infty} \frac{\partial}{\partial v_x}(a_x\chi f)\,dv_x dv_y dv_z = \iint_{-\infty}^{+\infty} dv_y dv_z\,[a_x\chi f]_{-\infty}^{+\infty} = 0 \qquad [2.12]$$

This cancels out because the distribution function $f$ is normalized and must therefore tend to zero when the speed tends to infinity. Thus:

$$\int_v \chi\vec{\gamma}\cdot\vec{\nabla}_v f d^3\vec{v} = -n\langle\vec{a}\cdot\vec{\nabla}v\chi\rangle \qquad [2.13]$$

Combining the previous results, we finally have:

$$\frac{\partial}{\partial t}(n\langle\chi\rangle) + \vec{\nabla}x\cdot(n\langle\chi\vec{v}\rangle) - n\left\langle\vec{a}\cdot\vec{\nabla}_v\chi\right\rangle = 0 \qquad [2.14]$$

This equation is called the *macroscopic transport equation*. It describes the evolution of the mean value of any quantity, depending on velocity, under the effect of mean forces $\vec{F} = \vec{a}/m$. For the moment, function $\chi$ has not been specified, and that is the whole point of this equation. Indeed, it is by choosing specific functions $\chi$ that we can establish hydrodynamic equations; these will be particular cases of the general transport equation [2.14].

## 2.3. Fluid model

First, to establish the conservation equation for the mass or the number of particles (equation [2.1]), let us choose $\chi(\vec{v}) = 1$. In this case, $\chi$ is not quite a function of velocity or, in other words, it is a polynomial in $\vec{v}$ of order 0. We therefore have:

$$\begin{aligned}
&\langle\chi\rangle = 1\\
&\langle\chi\vec{v}\rangle = \langle\vec{v}\rangle = \vec{u}\\
&\vec{\nabla}v\chi = \vec{\nabla}_v 1 = 0
\end{aligned} \qquad [2.15]$$

By substituting these results in the general transport equation [2.14], we get the *continuity equation*:

$$\frac{\partial n}{\partial t} + \vec{\nabla}\cdot(n\vec{u}) = 0 \qquad [2.16]$$

Equation [2.16] expresses the conservation of the number of particles. It can, in the same way, be expressed in terms of mass density. By choosing $\chi(\vec{v}) = m$, we have:

$$\begin{aligned} \langle \chi \rangle &= m \\ \langle \chi \vec{v} \rangle &= m \langle \vec{v} \rangle = m\vec{u} \\ \vec{\nabla}_v \chi &= \vec{\nabla}_v m = 0 \end{aligned} \qquad [2.17]$$

Substituting these results into the general transport equation [2.14] yields the continuity equation in terms of mass density:

$$\frac{\partial}{\partial t}(\rho) + \vec{\nabla} \cdot (\rho \vec{u}) = 0 \qquad [2.18]$$

where $\rho = mn$ is the density (mass per unit volume). We can easily pass from the equation of the conservation of the mass or the number of particles to the conservation of the charge (which can be interesting in the study of plasma), that is:

$$\frac{\partial}{\partial t}(\rho_\alpha) + \vec{\nabla} \cdot \left( \vec{J}_\alpha \right) = 0 \qquad [2.19]$$

where $\vec{J}_\alpha = q_\alpha \vec{u}_\alpha$ is the current, $q_\alpha$ is the charge and $\rho_\alpha$ is the charge density of the $\alpha$ species (electrons, ions, dust grains, etc.).

We can, in the same way, establish the relationship of conservation of momentum. For this, let us choose $\chi(\vec{v}) = m\vec{v}$. Thus, we have[1]:

$$m \int \mathbf{v} \frac{\partial f}{\partial t} d\mathbf{v} + m \int \mathbf{v}(\mathbf{v}.\nabla) f d\mathbf{v} + \int \mathbf{vF} \cdot \frac{\partial f}{\partial \mathbf{v}} d\mathbf{v} = 0 \qquad [2.20]$$

Again, let us go term by term. First, we can note that the first term can be written as follows:

$$m \int \mathbf{v} \frac{\partial f}{\partial t} d\mathbf{v} = m \frac{\partial}{\partial t} \int \mathbf{v} f d\mathbf{v} = m \frac{\partial}{\partial t} (n\mathbf{u}) \qquad [2.21]$$

Meanwhile, the third integral of relationship [2.20] becomes:

$$\begin{aligned} \int \mathbf{vF} \cdot \frac{\partial f}{\partial \mathbf{v}} d\mathbf{v} &= \int \frac{\partial}{\partial \mathbf{v}} \cdot [f \mathbf{vF} \cdot d\mathbf{v}] \\ &- \int f \mathbf{v} \frac{\partial}{\partial \mathbf{v}} \cdot \mathbf{F} d\mathbf{v} - \int f \mathbf{F} \cdot \frac{\partial}{\partial \mathbf{v}} \mathbf{v} \cdot d\mathbf{v} \end{aligned} \qquad [2.22]$$

1 To simplify the notations, we will simply denote vectors in bold. Also, the operator $\vec{\nabla}_v$ will be denoted as $\partial / \partial \mathbf{v}$.

The first two terms of relationship [2.22] cancel each other out for the same reasons as before. We also have:

$$\int \mathbf{vF} \cdot \frac{\partial f}{\partial \mathbf{v}} d\mathbf{v} = -\int \mathbf{F} f d\mathbf{v} \tag{2.23}$$

To evaluate the remaining term, we observe that velocity is a variable independent of space and time coordinates. Therefore, we have[2]:

$$\begin{aligned} \int \mathbf{v}(\mathbf{v}.\nabla) d\mathbf{v} &= \int \nabla \cdot (f\mathbf{vv}) d\mathbf{v} \\ &= \nabla \cdot \int f\mathbf{vv} d\mathbf{v} \\ &= \nabla \cdot n\langle \mathbf{vv} \rangle \end{aligned} \tag{2.24}$$

As discussed in section 1.4, it can be useful to separate velocity into a mean part $\mathbf{u}$ (or fluid velocity) and a thermal part $\mathbf{c}$ (see equation [1.74]). Thus:

$$\nabla \cdot n\langle \mathbf{vv} \rangle = \nabla \cdot (n\mathbf{uu}) + \nabla \cdot (n\langle \mathbf{cc} \rangle) + 2\nabla \cdot (n(\mathbf{u}\langle \mathbf{c} \rangle)) \tag{2.25}$$

By combining the previous terms and using the continuity equation for the first two terms, we get[3]:

$$m_\alpha n_\alpha \left[ \frac{\partial \mathbf{u}_\alpha}{\partial t} + (\mathbf{u}_\alpha \cdot \nabla) \mathbf{u}_\alpha \right] = n_\alpha \langle \mathbf{F} \rangle - \nabla \cdot \overleftrightarrow{p}_\alpha \tag{2.26}$$

In particular, in the case of a plasma, we have:

$$\langle \mathbf{F} \rangle = q_\alpha (\mathbf{E} + \mathbf{u}_\alpha \times \mathbf{B}) \tag{2.27}$$

while for a system of self-gravitating particles, we have:

$$\langle \mathbf{F} \rangle = m_\alpha \mathbf{g} \tag{2.28}$$

where $\mathbf{E}$, $\mathbf{B}$ and $\mathbf{g}$ represent mean fields. Index $\alpha$ has been added to specify that in the presence of several species (as in the case of a plasma), each species must be treated separately. Equation [2.26] is called the *Euler equation* or conservation of momentum

2 Note that products involving the velocity vector $\mathbf{vv}$ are understood in the sense of a (dyadic) tensor product. With a more explicit notation, we will denote it as $\mathbf{v} \otimes \mathbf{v}$. This is the reason why pressure will be interpreted as a tensor (see section 2.4).

3 We will denote matrix quantities as $\overleftrightarrow{\bullet}$ to distinguish them from vectors. Thus, in equation [2.26], the pressure tensor $\overleftrightarrow{p_\alpha}$ represents a $3 \times 3$ matrix.

equation. It translates the variation of an element of the fluid under the effect of the average force and the effects of pressure and viscosity, both contained in the last term $\overleftrightarrow{p}_\alpha$, the *kinetic pressure tensor*, defined as follows:

$$\overleftrightarrow{p}_\alpha(\mathbf{r},t) \equiv n_\alpha m_\alpha \langle \mathbf{cc} \rangle = m_\alpha \int f\mathbf{cc}d\mathbf{c} \qquad [2.29]$$

where $\mathbf{c} = \mathbf{v} - \mathbf{u}_\alpha$ is the thermal component of the velocity, referred to as central moments.

Note that the previous reasoning can very well be extended to higher orders. Thus, the evolution of the energy flow is obtained by taking the next moment of the Vlasov equation, that is, for $\chi \equiv m\mathbf{vv}/2$, and by integrating on velocity. However, we will not discuss this aspect further here. We will come back to this point in section 2.5.

## 2.4. Pressure tensor

Equations describing the evolution of moments inevitably seem to get more complicated as we consider higher order moments. Thus, the equation describing the evolution of density [2.16], which is scalar, involves fluid velocity, therefore, a vector. Similarly, the equation describing the variation of the fluid velocity vector or the momentum (equation [2.26]) involves the pressure tensor, that is, a $3 \times 3$ matrix. This is because the product is understood in the broad sense of a tensor product (dyadic product). The pressure cannot therefore be assimilated to a scalar without justification. In fact, the thermodynamic approximation a graduate student should be familiar with, where pressure is represented by a scalar, is only justified in the presence of collisions. Thus, it is important here to redefine in physical terms what is hidden, more precisely, under the term "pressure".

Pressure is generally defined as the force per unit area exerted by a gas (by collision effect) on the walls that surround it. If we take an elementary surface $d\mathbf{s}$, pressure simply represents the transfer of momentum per unit area and time. Pressure is therefore the flow of momentum on surface $d\mathbf{s}$, due to the random movement of molecules $\vec{c}$ which make up the gas. Pressure therefore has the form $\overleftrightarrow{p} = \rho\langle \mathbf{c} \otimes \mathbf{c} \rangle$, where $\mathbf{c}$ represents the random or thermal part of the velocity.

Thus, pressure only involves random velocities compared to the mean velocity of fluid $\mathbf{u}$, that is, $\langle (\mathbf{v} - \mathbf{u}) \otimes (\mathbf{v} - \mathbf{u}) \rangle$. We refer to *central* moments. Here, we will not discuss the subtleties related to the pressure tensor, which will not be used in Part 2 of this book, which is devoted to applications. It is, however, important to give the most usual approximations of the pressure tensor and their physical significance.

Note that the pressure tensor, being a matrix $(3 \times 3)$, can be separated into a diagonal component (representing the pressure in the usual sense) and off-diagonal components (representing what is called *viscosity*):

$$\overleftrightarrow{p} = p\overleftrightarrow{\delta} - \overleftrightarrow{\pi} \quad [2.30]$$

where $\overleftrightarrow{\delta}$ is the diagonal unit tensor and $\overleftrightarrow{\pi}$ is the viscosity tensor, grouping only the terms outside the diagonal.

In the case of an isotropic distribution, momentum exchanges are always parallel to the surface $d\mathbf{s}$ through which the flow passes. In this case, we can write $\overleftrightarrow{p} \cdot d\mathbf{s} = pd\mathbf{s}$ and the pressure tensor is reduced to:

$$\overleftrightarrow{p} = p\overleftrightarrow{I} \quad [2.31]$$

$\overleftrightarrow{I}$ being the unit tensor. The pressure tensor is therefore reduced in this case to the usual thermodynamic definition of a scalar pressure $p$.

In the case of a collision-free system, however, this definition is not justified. We will now look at the two most commonly used types of pressure in fluid equations.

### 2.4.1. *Isotropic pressure*

An initial approximation is to assume that the system is isotropic. So, we have:

$$\overleftrightarrow{p} = \begin{pmatrix} p & 0 & 0 \\ 0 & p & 0 \\ 0 & 0 & p \end{pmatrix} \quad [2.32]$$

Although this hypothesis is difficult to justify in the case of a plasma (where the presence of magnetic fields essentially tends to make the medium anisotropic), experience shows that distribution functions that are too anisotropic are unstable. Thus, this approximation remains, to a certain extent, applicable.

### 2.4.2. *Anisotropic pressure*

In the case of a magnetized plasma where the presence of magnetic fields renders the system anisotropic, a diagonal pressure tensor can be used in the frame of reference linked to the magnetic field. It will be considered that the pressure on the perpendicular plane is the same in both directions. We will then write:

$$\overleftrightarrow{p} = \begin{pmatrix} p_\perp & 0 & 0 \\ 0 & p_\perp & 0 \\ 0 & 0 & p_\parallel \end{pmatrix} \qquad [2.33]$$

where it has been assumed that the magnetic field is parallel to the $z$-axis. Here, $p_\perp$ represents the pressure on the plane perpendicular to the magnetic field and $p_\parallel$ represents the pressure on the axis parallel to the field.

## 2.5. A deadlock in the fluid model: closure relationships

The problem with the fluid model is that it does not constitute a closed system of equations. Thus, the equation of continuity, which expresses the density variation, involves a new unknown, that is to say the fluid velocity. Similarly, the equation that determines its evolution, namely the Euler equation, involves the pressure term. We can try to go further and establish equations governing the evolution of higher moments, but this only shifts the problem to higher orders. Thus, the equation governing the evolution of the moment of order $n$ will involve the moment of order $n+1$. Thus, these equations become more and more complicated, and difficult to handle in practice, because they involve tensors of increasing order.

It is necessary to close the system of equations to overcome this problem of infinite hierarchy. Closure relationships consist of truncating the fluid model equations, either on the order 2 (i.e. on pressure), or on the moment of order 3, that is, the heat flow. Here, we will focus only on the truncations on the moment of order 2.

Two commonly used approximations make it possible to make the hydrodynamic model a closed system by introducing approximations on pressure. Here, we will mention the two most common approximations (Raimbault and Smets 2014).

### 2.5.1. *Zero pressure*

In situations where the pressure within the fluid may be ignored in relation to other contributions, the following are written (in the isotropic case):

$$p \equiv 0 \qquad [2.34]$$

The system of equations to be considered then simply becomes:

$$\frac{\partial n}{\partial t} + \nabla \cdot (n\mathbf{u}) = 0$$
$$mn\left[\frac{\partial \mathbf{u}}{\partial t} + (\mathbf{u} \cdot \nabla)\,\mathbf{u}\right] = n\langle \mathbf{F} \rangle \qquad [2.35]$$

In this case, the Euler equation now involves only the moment of order 1 (mean velocity) and the hydrodynamic system, with the equations describing the fields (Maxwell equations in the case of a plasma or the Poisson equation of the gravitational field), constitutes a closed system of equations. In the case of plasmas, this approximation is known as the *cold plasma approximation*, because it amounts to considering an equation of state $p = nk_BT$ for a zero temperature $T = 0$.

### 2.5.2. *Isothermal pressure*

Another approximation consists of assuming that each component of the system (electrons and ions in the case of a plasma) satisfies the equation of state of ideal gases, that is,

$$p = nk_BT \qquad [2.36]$$

In the case where the temperature varies only very little between one point and another in the medium, it can be considered constant. In this case, the gradient of the pressure, appearing in the Euler equation [2.26], only involves the gradient of the density, namely:

$$\nabla p = k_BT\nabla n \qquad [2.37]$$

This relationship also makes it possible to close the system of equations of the model, provided, however, that the temperature is assumed to be known. Otherwise, this relationship would only be a defining relationship of a new variable, temperature, from pressure and density.

There are, however, truncations that relate to the moment of higher order (order 3); this is called the heat flux tensor $\overleftrightarrow{Q}$. It will generally be considered that the heat flux is constant, that is $\vec{\nabla} \cdot \overleftrightarrow{Q} = 0$. This simply translates the *adiabatic hypothesis*.

In plasmas, this hypothesis is well justified in general. Indeed, the adiabatic hypothesis is verified in the general case of perturbations that propagate parallel to the magnetic field lines and at high velocities, with respect to the velocity of particles. This approximation is justified in the case of high thermal conductivity, where temperature gradients are compensated by conduction effects.

## 2.6. The collision effect

So far, we have considered the absence of collisions. However, the effect of collisions can be introduced into the fluid approach. To do this, we can follow the same steps described in sections 2.2 and 2.3, but take the collision term of the Boltzmann equation [1.24] into account. Multiplying the Boltzmann equation by a function of the velocity $\chi(\mathbf{v})$, and integrating the velocities over space, we have:

$$\int \chi(\mathbf{v}) \left( \frac{\partial f}{\partial t} + \mathbf{v} \cdot \frac{\partial f}{d\mathbf{r}} + \mathbf{a} \frac{\partial f}{\partial \mathbf{v}} \right) d\mathbf{v} = \int \chi(\mathbf{v}) \left( \frac{\delta f}{\delta t} \right)_{coll} d\mathbf{v} \qquad [2.38]$$

By observing that position and time are velocity-independent variables, the first two terms can be written as follows:

$$\int \chi(\mathbf{v}) \left( \frac{\partial f}{\partial t} + \mathbf{v} \cdot \frac{\partial f}{\partial \mathbf{r}} \right) d\mathbf{v} = \frac{\partial \langle n\chi(\mathbf{v}) \rangle}{\partial t} + \frac{\partial \langle n\chi(\mathbf{v})\mathbf{v} \rangle}{\partial \mathbf{r}} \qquad [2.39]$$

where we used the definition of the mean value. Furthermore, assuming, as in section 2.2 that the component of the force field along an axis $i$ does not depend on velocity along this axis, we have:

$$\begin{aligned} \int \chi(\mathbf{v}) \mathbf{a} \cdot \frac{\partial f}{\partial \mathbf{v}} d\mathbf{v} &= \int \chi(\mathbf{v}) \frac{\partial (\mathbf{a}f)}{\partial \mathbf{v}} d\mathbf{v} = -\int \frac{\partial \chi(\mathbf{v})}{\partial \mathbf{v}} \cdot \mathbf{a} f d\mathbf{v} \\ &= -\left\langle n\mathbf{a} \cdot \frac{\partial \chi(\mathbf{v})}{\partial \mathbf{v}} \right\rangle \end{aligned} \qquad [2.40]$$

where, as in section 2.2, we have done one integration per part and taken into account the fact that the distribution is normalized. The amount due to collisions is defined as follows:

$$\left( \frac{\delta \langle n\chi(\mathbf{v}) \rangle}{\delta t} \right)_c \equiv \int \chi(\mathbf{v}) \left( \frac{\delta f}{\delta t} \right)_{coll} d\mathbf{v} \qquad [2.41]$$

The general transport equation is written as:

$$\frac{\partial \langle n\chi(\mathbf{v}) \rangle}{\partial t} + \nabla \cdot (\langle n\chi(\mathbf{v})\mathbf{v} \rangle) = \left\langle \mathbf{a} \cdot \frac{\partial n\chi(\mathbf{v})}{\partial \mathbf{v}} \right\rangle + \left( \frac{\delta \langle n\chi(\mathbf{v}) \rangle}{\delta t} \right)_{coll} \qquad [2.42]$$

To obtain the continuity equation and the Euler equation in this case, it is enough to identify the function $\chi(\mathbf{v})$ at a velocity power. Particularly for $\chi(\mathbf{v}) = m$ and $\chi(\mathbf{v}) = m\mathbf{v}$, we obtain the conservation of mass and the conservation of momentum as follows:

$$\begin{aligned} &\frac{\partial(nm)}{\partial t} + \nabla \cdot (nm\langle \mathbf{v}\rangle) = \left(\frac{\delta\langle nm\rangle}{\delta t}\right)_{coll} \\ &\frac{\partial(nm\langle \mathbf{v}\rangle)}{\partial t} + \nabla \cdot (nm\langle \mathbf{vv}\rangle) - n\langle \mathbf{F}\rangle = \left(\frac{\delta(nm\langle \mathbf{v}\rangle)}{\delta t}\right)_{coll} \end{aligned} \qquad [2.43]$$

In the absence of collisions, the right-hand side of the previous equation system cancels each other out and we find the equations of the collisionless hydrodynamic model, as discussed in section 2.3. The effect of collisions appears as a source term of equations representing the variation in density and momentum due to collisions. Equations [2.43] can be rewritten in the following more compact form:

$$\begin{aligned} &\frac{\partial \rho}{\partial t} + \nabla \cdot \Gamma = \left(\frac{\delta \rho}{\delta t}\right)_{coll} \\ &\frac{\partial \Gamma}{\partial t} + \nabla . \Sigma = n\langle \mathbf{F}\rangle + \left(\frac{\delta \Gamma}{\delta t}\right)_{coll} \end{aligned} \qquad [2.44]$$

where $\rho$, $\Gamma$ and $\Sigma$ are the first three moments of the distribution, namely:

$$\begin{aligned} &\rho(\mathbf{r},t) \equiv nm\langle 1\rangle = m\int f d\mathbf{v} \\ &\Gamma(\mathbf{r},t) \equiv nm\langle \mathbf{v}\rangle = m\int f\mathbf{v}d\mathbf{v} \\ &\Sigma(\mathbf{r},t) \equiv nm\langle \mathbf{vv}\rangle = m\int f\mathbf{vv}d\mathbf{v} \end{aligned} \qquad [2.45]$$

Equations [2.44] can also be written in the most useful form:

$$\begin{aligned} &\frac{\partial n}{\partial t} + \nabla \cdot (n\mathbf{u}) = \left(\frac{\delta n}{\delta t}\right)_{coll} \\ &nm\left(\frac{\partial \mathbf{u}}{\partial t} + \mathbf{u} \cdot \nabla \mathbf{u}\right) + \nabla \cdot \overleftrightarrow{P} - n\langle \mathbf{F}\rangle = \left(\frac{\delta \mathbf{M}}{\delta t}\right)_{coll} \end{aligned} \qquad [2.46]$$

where the source term of the continuity equation is given by:

$$\left(\frac{\delta n}{\delta t}\right)_{coll} = \int \left(\frac{\delta f}{\delta t}\right)_{coll} d\mathbf{v} \equiv S \qquad [2.47]$$

and the source term of the Euler equation is written as:

$$\left(\frac{\delta \mathbf{M}}{\delta t}\right)_{coll} = \mathbf{A} - Sm\mathbf{u} \qquad [2.48]$$

with:

$$\mathbf{A} \equiv \int m\mathbf{v}\left(\frac{\delta f}{\delta t}\right)_{coll} d\mathbf{v} \qquad [2.49]$$

# 3

# Quantum Models

*Nobody exists on purpose.*
*Nobody belongs anywhere.*
*Everybody's gonna die.*
*Come watch TV.*

Rick Sanchez (*Rick & Morty*)

In the previous two chapters, we have seen two statistical descriptions of plasmas, and more generally, of complex systems where collective phenomena take place. These two approaches have been developed in a classic framework, that is to say without taking into account principles emanating from quantum mechanics. So, it is natural to ask the following questions: *does such a description necessarily have to remain classic or is it possible to extend it to the framework of quantum mechanics?* We will see in this chapter that such an extension is indeed possible.

Beyond the purely theoretical interest that such a generalization to quantum mechanics can have, there are practical motivations for this question. First, in plasmas, the intuitive image of a plasma as a hot gas, where the effects of temperature make quantum phenomena insignificant, is not correct. Modern technologies make it possible to produce cold and extremely dense plasmas. This ushered in a new era in plasma physics, that of *quantum plasmas*. Moreover, there are astrophysical objects where matter is so dense, for example, neutron stars, that collective phenomena that operate there must be described in a quantum context. Moreover, these collective phenomena, and the statistical approaches to understanding them, are not confined to plasma physics, and a quantum statistical description is needed to understand other physical problems, such as Bose–Einstein condensates, cold atoms in optical gratings or certain dark matter models. As we will see in Part 2, devoted to applications, these systems are also the site of collective phenomena similar to those studied in plasmas and can be described by formally similar approaches.

## 3.1. Physical interest

Quantum effects are important when matter is dense and the temperature is low. To understand why it happens, it is necessary to explain what is behind the term "quantum effects". In reality, these are two distinct, albeit interrelated, effects. The first is the effect of *quantum statistics*. Thus, electrons in a plasma at equilibrium, for example, do not entirely obey the Boltzmann distribution [1.56], due to the Pauli exclusion principle which prevents a large number of electrons from occupying the same energy level. As electrons are fermions, they obey the Fermi–Dirac statistics, whose Boltzmann distribution is a good approximation when the temperature is high ("high" with respect to the Fermi temperature). This effect is therefore significant when the temperature is low.

The second effect is a purely dynamic effect, that is, related to the displacement of the particles in the medium. It is this effect that will interest us the most if we want to generalize hydrodynamic and kinetic models to the quantum mechanics context. To understand its origin, we must remember that fundamentally, at least on a microscopic scale, particles have a wave-like behavior. According to the standard interpretation of quantum mechanics, known as the *Copenhagen interpretation*, there is no need to talk about the trajectory of an elementary particle (an electron, for example). In other interpretations of quantum mechanics, such as in the *de Broglie–Bohm* theory, it is however possible to associate trajectories with elementary particles, but these trajectories are highly non-classic. In fact, however, at the microscopic scale, it is the wave functions, and the Schrödinger equation governing their evolution, that are fundamental. This wave character induces a certain number of phenomena for which there is no classical analog, such as quantum diffraction, interference phenomena and the tunnel effect. This behavior is significant only at small spatial scales and is felt only when the mean particle distance is comparable to the *de Broglie wavelength*. This effect is therefore significant when the material is dense.

As an example, Figure 3.1 shows a classification of plasmas. The continuous straight line segment shows the boundary between classic plasmas (high temperature and low density) and quantum plasmas (low temperature and high density). Table 3.1 also shows the typical density and temperature values of different types of natural or laboratory plasmas. Quantum plasmas include electrons in a metal or dense astrophysical objects, such as white dwarfs.

The two approaches, kinetic and hydrodynamic, presented in Chapters 1 and 2, can be extended to the quantum mechanics context. We will start by presenting the

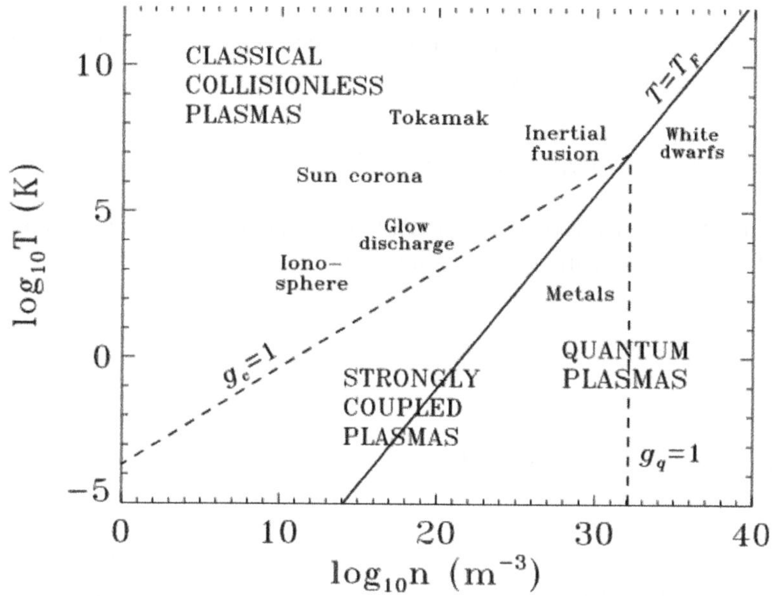


**Figure 3.1.** *Diagram of phases (logarithmic scale), showing the different types of plasmas. The continuous straight line segment delimits the classical plasmas of quantum plasmas (source: Manfredi et al. (2019))*

| | $n(m^{-3})$ | T(eV) |
|---|---|---|
| **Interstellar plasma** | $10^6$ | $10^{-2}$ |
| **Solar chromosphere** | $10^{18}$ | 2 |
| **Solar wind** | $10^7$ | 10 |
| **Ionosphere** | $10^{12}$ | 0,1 |
| **Discharge arc** | $10^{20}$ | 1 |
| **Tokamak** | $10^{20}$ | $10^4$ |
| **Stars** | $10^{26}$ | $2 \times 10^3$ |
| **Inertial confinement fusion** | $10^{28}$ | $10^4$ |

**Table 3.1.** *Density and temperature parameter (electronics) for a variety of natural and laboratory plasmas*

hydrodynamic approach because its generalization to quantum mechanics is quite simple. We will then present the quantum generalization of kinetic theory that requires the predefinition of a certain number of concepts.

## 3.2. Quantum hydrodynamic model

To generalize the hydrodynamic model to the quantum mechanics context, let us take the Schrödinger equation as a starting point for our $N$-particle system:

$$i\hbar\frac{\partial\Psi_N}{\partial t} = \hat{H}_N\Psi_N \qquad [3.1]$$

where

$$\hat{H}_N = \sum_{i=1}^{N}\left\{-\frac{\hbar^2}{2m}\nabla_i^2 + V\left(\mathbf{r}_i\right)\right\} \qquad [3.2]$$

is the Hamiltonian of the system and $\Psi_N$ is the $N$-particle wave function. Such a model is clearly insoluble for the same reasons discussed in the conventional context, namely the extremely large number of particles making up the systems in question, typically $N \sim 10^{23}$. Also, a macroscopic model will involve individual Schrödinger equations with a single particle, namely:

$$i\hbar\frac{\partial\psi_\alpha}{\partial t} = -\frac{\hbar^2}{2m}\nabla^2\psi_\alpha + V\psi_\alpha \qquad [3.3]$$

where we introduced the index $\alpha$ to note that, in the case of a system composed of several species of particles like a plasma, each species will have to be treated separately. In the case of a plasma, it will typically be electrons and ions and, possibly, neutral atoms, if the plasma is only partially ionized. In equation [3.3], potential $V$ represents a mean potential, that is the average potential generated by the set of particles that make up the medium. The hydrodynamic representation of Schrödinger's equation is actually almost as old as Schrödinger's equation itself. We owe it mainly to the German mathematician *Erwin Madelung* (1927), who noticed that Schrödinger's equation could be rewritten formally as a fluid model. To show this, let us start by writing the wave function in polar form:

$$\psi_\alpha(\mathbf{r},t) = A_\alpha(\mathbf{r},t)\exp\left(iS_\alpha(\mathbf{r},t)/\hbar\right) \qquad [3.4]$$

Since the wave function is a complex function, it can always be written in this form. In equation [3.4], $A_\alpha(r,t)$ and $S_\alpha(r,t)$ are real functions, representing the amplitude and phase of the wave function. By replacing the wave function in its polar form [3.4] in the Schrödinger equation [3.3], and separating the real and imaginary parts of the equation, we have:

$$\partial_t S_\alpha + \frac{1}{2m}(\nabla S_\alpha)^2 + V(\mathbf{r}) + \frac{\hbar^2}{2m}\frac{\nabla^2 A_\alpha}{A_\alpha} = 0$$
$$\partial_t A_\alpha + \frac{1}{2m}(\nabla S)\cdot\nabla A_\alpha + A_\alpha \nabla^2 S_\alpha = 0 \quad [3.5]$$

It may be interesting to note that the reduced Planck constant $\hbar$ appears only in the real part of the Schrödinger equation, through the last term of the first equation of the system [3.5]. The imaginary part, on the other hand, is a classic equation, not involving the Planck constant. We can note that in the $\hbar \to 0$ limit (classical limit), the first equation of the system [3.5] is reduced to:

$$\partial_t S_\alpha = -\left[\frac{1}{2m}(\nabla S_\alpha)^2 + V\right] \quad [3.6]$$

It is indeed a well-known equation of classic Hamiltonian mechanics, namely the *Hamilton–Jacobi* equation. By comparing this last equation [3.6] with the first equation of the system [3.5], it appears that the difference between classic dynamics and quantum dynamics is contained in the term proportional to $\hbar^2$. We can easily verify that this term has the dimension of a potential (or of an energy).

To obtain hydrodynamic equations, we need to define a density and a fluid velocity. Let us define density as simply the modulus squared of the amplitude, namely:

$$n_\alpha \equiv |\psi_\alpha|^2 = A_\alpha^2 \quad [3.7]$$

Let us also define the fluid velocity as follows:

$$\mathbf{u}_\alpha = \frac{1}{m}\nabla S_\alpha \quad [3.8]$$

The definition of [3.7] of density is quite intuitive: it simply reflects the fact that the modulus squared of the wave function represents a probability density of presence. The definition of fluid velocity [3.8] is a little less obvious. A dimensional analysis easily shows that $\mathbf{u}_\alpha$, thus defined, does represent a velocity, but why should it be perceived as a fluid velocity? The reason for this is simple; this definition retains the definition of current, namely:

$$\mathbf{j}_\alpha = n_\alpha \mathbf{u}_\alpha = \frac{\hbar}{2m_\alpha i}\left[\psi_\alpha^*(\nabla\psi_\alpha) - \psi_\alpha(\nabla\psi_\alpha^*)\right] \quad [3.9]$$

Using transformations [3.7] and [3.8], equations [3.5] become:

$$\begin{aligned} &\frac{\partial n_\alpha}{\partial t} + \nabla\left(n_\alpha \mathbf{u}_\alpha\right) = 0 \\ &\frac{\partial \mathbf{u}_\alpha}{\partial t} + \mathbf{u}_\alpha \nabla \mathbf{u}_\alpha = -\frac{\nabla V}{m_\alpha} - \frac{\hbar^2}{2m_\alpha^2}\nabla\left(\frac{\nabla^2\sqrt{n_\alpha}}{\sqrt{n_\alpha}}\right) \end{aligned} \qquad [3.10]$$

where we also applied the gradient $\nabla$ to the second equation of the system [3.5].

Equations [3.10] form the *quantum hydrodynamic* model. An analysis of the equations [3.10] reveals a number of interesting points: first, note that the first equation is none other than the continuity equation [2.1], obtained in the previous chapter. It is not affected by the presence of quantum effects. It is also interesting to note that it is contained in the Schrödinger equation. In fact, this translates the notion of conservation of probability in quantum mechanics. The second equation of the system [3.10] shows the emergence of a new term, with the dimension of a potential:

$$Q(\mathbf{r},t) = -\frac{\hbar^2}{2m}\frac{\nabla^2 A_\alpha}{A_\alpha} = -\frac{\hbar^2}{2m}\frac{\nabla^2\sqrt{n_\alpha}}{\sqrt{n_\alpha}} \qquad [3.11]$$

This term is called the *quantum potential* or *Bohm potential*. Indeed, it is this term that reflects the wave effects, namely the effects of diffraction and interference. Note that this potential can also be represented in terms of force. Thus, we have a force deriving from this potential given by:

$$\mathbf{F_Q} = -\nabla Q(\mathbf{r},t) \qquad [3.12]$$

This force is usually called the *quantum pressure force*.

Bohm's quantum potential has a number of non-trivial properties that would be interesting to discuss here. Thus, Hiley (1999, 2002) points out the following aspects:

– The Bohm potential derives from the real part of the Schrödinger equation and not from the Hamiltonian or any other external source.

– If we multiplied the amplitude of the wave function $A_\alpha$ by a constant, the Bohm potential [3.11] would remain invariant (because the amplitude appears in both the numerator and the denominator). It is thus independent of the amplitude. It is then said to have a non-local character because its effect does not decrease with distance, unlike ordinary potentials, such as the Coulomb potential or the gravitational potential.

– It contains information on the system as a whole. Thus, in an experiment in quantum mechanics, the Bohm potential contains information about the entire experimental device where the particle is located.

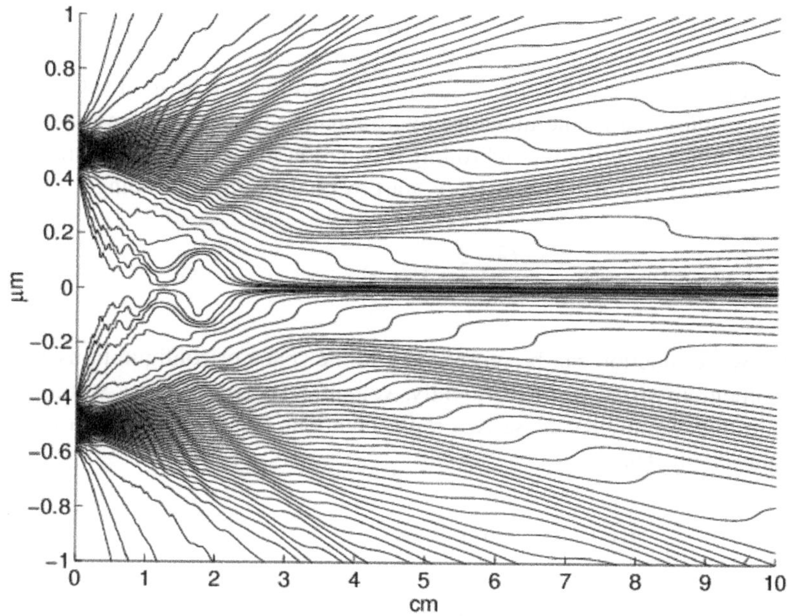


**Figure 3.2.** *Young's slit experiment with electrons*

COMMENT ON FIGURE 3.2.– *The figure shows the trajectories predicted by the de Broglie–Bohm theory, where the Bohm potential gives non-classical trajectories to electrons. The probability density calculated by the de Broglie–Bohm theory is exactly the same as that calculated by usual quantum mechanics and reproduces the observations very well (source: Gondran and Gondran (2014)).*

It may be useful to start a side note to clarify that there is an interpretation of quantum mechanics (the *de Broglie–Bohm* theory (Bricmont 2021)), where this term represents the fundamental difference between classic physics and quantum mechanics. Thus, even in a vacuum, according to de Broglie–Bohm's interpretation, particles do not follow rectilinear trajectories as in classic physics, but their movement is "guided" by quantum potential [3.11]. It is in this way that this interpretation explains, for example, the experience of the dual slits (Young's slits) where electrons produce an interference pattern. We show this in Figure 3.2, representing a numerical simulation of trajectories in the presence of the Bohm quantum potential.

## 3.3. Quantum kinetic approach: Wigner–Moyal theory

In section 3.2, we saw that the Schrödinger equation can be reduced to a system of fluid equations, that is, the quantum hydrodynamic model. Here, we will see that it is also possible to rewrite the Schrödinger equation in a kinetic form similar to the Vlasov equation [1.25]. To do this, let us start by considering Schrödinger's equation:

$$i\hbar\frac{\partial}{\partial t}\psi = H\psi, \quad H = -\frac{\hbar^2}{2m}\nabla^2 + V(\mathbf{r},t) \qquad [3.13]$$

where the wave function can be represented in the space of configurations $\psi(\mathbf{r},t)$ or in the momentum space (or, equivalently, velocities) $\psi(\mathbf{p},t)$. It is a standard result in quantum mechanics that these two representations are related to each other by a Fourier transform. That is:

$$\psi(\mathbf{p},t) = \int \psi(\mathbf{r},t)\exp(-i\mathbf{p}\cdot\mathbf{r}/\hbar)d\mathbf{r} \qquad [3.14]$$

Thus, the main problem we face when looking for a kinetic representation of the Schrödinger equation is to represent the probabilities in the *phase space*. Now, in quantum mechanics, the probabilities (related to the position or to the quantity of movement) are known and are in fact given by the modulus squared of the wave functions. Indeed, the probability of finding the particle around the point $\mathbf{r}$ (at the instant $t$) is given by $P(\mathbf{r},t) = |\psi(\mathbf{r},t)|^2 d\mathbf{r}$. Similarly, the probability in the momentum space is given by $P(\mathbf{p},t) = |\psi(\mathbf{p},t)|^2 d\mathbf{p}$. Since both probabilities are normalized, we have:

$$\int P(\mathbf{r},t)d\mathbf{r} = \int P(\mathbf{p},t)d\mathbf{p} = 1 \qquad [3.15]$$

Thus, to construct a quantum kinetic theory, it will be necessary to combine in a certain way both the information contained in $P(\mathbf{r},t)$ and that contained in $P(\mathbf{p},t)$ into a single distribution, defined on the phase space $(\mathbf{r},\mathbf{p})$.

### 3.3.1. *Wigner distribution*

Since the kinetic approach is formulated in phase space, it is necessary to combine these two probabilities $P(\mathbf{r},t)$ and $P(\mathbf{p},t)$ into a single expression: the result will represent the quantum equivalent of the distribution function. Such a definition was proposed by *Eugene Wigner* (1932). The *Wigner distribution*[1] is defined as follows:

$$W(\mathbf{r},\mathbf{p},t) = \int \psi^*(\mathbf{r}+\mathbf{s}/2,t)\cdot\psi(\mathbf{r}-\mathbf{s}/2,t)\exp(-i\mathbf{p}\cdot\mathbf{s}/\hbar)d\mathbf{s} \qquad [3.16]$$

It is interesting to note that the Wigner distribution [3.16] does indeed contain information on both $P(\mathbf{r},t)$ and $P(\mathbf{p},t)$. Indeed, by integrating the Wigner distribution, we can show that we have:

$$P(\mathbf{r},t) = \int W(\mathbf{r},\mathbf{p},t)d\mathbf{p}, \quad P(\mathbf{p},t) = \int W(\mathbf{r},\mathbf{p},t)d\mathbf{r} \qquad [3.17]$$

Thus, *marginal* distributions (i.e. obtained by integrating on one of the variables, **r** or **p**) correspond well to the probabilities predicted by quantum mechanics, namely $P(\mathbf{r},t)$ and $P(\mathbf{p},t)$. In addition, it may be interesting to note that by defining:

$$K(\mathbf{r}_1,\mathbf{r}_2,t) := \psi(\mathbf{r}_1,t)\psi*(\mathbf{r}_2,t) \qquad [3.18]$$

we can express it as:

$$W(\mathbf{r},\mathbf{p},t) = \int K(\mathbf{r},\mathbf{s},t)\exp(-i\mathbf{p}\cdot\mathbf{s}/\hbar)d\mathbf{s} \qquad [3.19]$$

Thus, the Wigner distribution is none other than the Fourier transform of $K(\mathbf{r}_1,\mathbf{r}_2,t)$, where $\mathbf{r} = (\mathbf{r}_1+\mathbf{r}_2)/2$ and $\mathbf{s} = \mathbf{r}_1 - \mathbf{r}_2$. It represents the Fourier transform of the autocorrelation function corresponding to the wave function $\psi$. The Wigner distribution, thus defined, represents a form of probability of a quantum system in phase space. In this sense, it allows a link between quantum mechanics and statistical physics.

To study the properties of the Wigner distribution and establish the equation it obeys, let us consider a pure state $|\ \psi\rangle$, and let us introduce two closing relations, in the space of positions and in the momentum space, that is to say:

$$\int |\mathbf{r}\rangle\langle\mathbf{r}|d\mathbf{r} = 1, \quad \int |\mathbf{k}\rangle\langle\mathbf{k}|\frac{d\mathbf{k}}{(2\pi)^3} = 1 \qquad [3.20]$$

1 To be precise, we should actually speak of quasi-distribution rather than distribution because, although the Wigner function has real values, it can in some cases take negative values. However, these negative values are only present at small scales, below $\hbar$, when the phase space representation reaches its limits.

where $\mathbf{r}$ is the position operator and $\mathbf{p} = \hbar\mathbf{k}$ is the momentum operator. Thus, we have:

$$\langle \psi \mid \psi \rangle = \int \langle \psi \mid \mathbf{r} \rangle \langle \mathbf{r} \mid \psi \rangle d\mathbf{r} = \int \langle \psi \mid \mathbf{k} \rangle \langle \mathbf{k} \mid \psi \rangle \frac{d\mathbf{k}}{(2\pi)^3} = 1 \quad [3.21]$$

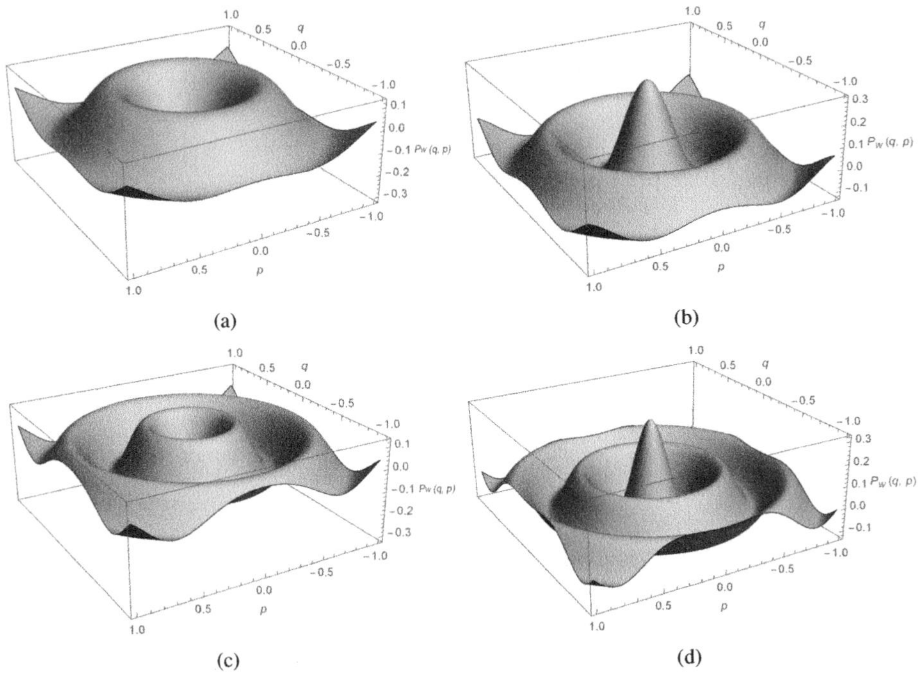


**Figure 3.3.** *Examples of Wigner distributions for a quantum harmonic oscillator for different energy levels. (a) n = 5; (b) n = 10; (c) n = 15; (d) n = 20. Note the appearance of negative values. For a color version of this figure, see www.iste.co.uk/ourabah/collective.zip*

Using the fact that:

$$\psi(\mathbf{r},t) = \langle \mathbf{r} \mid \psi \rangle \text{ and } \psi(\mathbf{k},t) = \langle \mathbf{k} \mid \psi \rangle \quad [3.22]$$

we also have:

$$\langle \psi \mid \psi \rangle = \int \psi^*(\mathbf{r},t)\psi(\mathbf{r},t)d\mathbf{r} = \int \psi^*(\mathbf{k},t)\psi(\mathbf{k},t)\frac{d\mathbf{k}}{(2\pi)^3} = 1 \quad [3.23]$$

By introducing two other closure relationships into relationships [3.21] and [3.23], we obtain:

$$\langle \psi \mid \psi \rangle = \int \frac{d\mathbf{k}}{(2\pi)^3} \int d\mathbf{r}' \int d\mathbf{r}'' \langle \psi \mid \mathbf{r}' \rangle \langle \mathbf{r}' \mid \mathbf{k} \rangle \langle \mathbf{k} \mid \mathbf{r}'' \rangle \langle \mathbf{r}'' \mid \psi \rangle = 1 \quad [3.24]$$

and:

$$\langle \psi \mid \psi \rangle = \int d\mathbf{r} \int \frac{d\mathbf{k}'}{(2\pi)^3} \int \frac{d\mathbf{k}''}{(2\pi)^3} \langle \psi \mid \mathbf{k}' \rangle \langle \mathbf{k}' \mid \mathbf{r} \rangle \langle \mathbf{r} \mid \mathbf{k}'' \rangle \langle \mathbf{k}'' \mid \psi \rangle = 1 \quad [3.25]$$

Let us now recall an elementary result of quantum mechanics, namely:

$$\langle \mathbf{r} \mid \mathbf{k} \rangle = \langle \mathbf{k} \mid \mathbf{r} \rangle^* = \exp(-i\mathbf{k} \cdot \mathbf{r}) \quad [3.26]$$

This implies that equations [3.24] and [3.25] can be rewritten as follows:

$$\langle \psi \mid \psi \rangle = \int \frac{d\mathbf{k}}{(2\pi)^3} \int d\mathbf{r}' \int d\mathbf{r}'' \langle \psi \mid \mathbf{r}' \rangle \langle \mathbf{r}'' \mid \psi \rangle \exp\left[-i\mathbf{k} \cdot (\mathbf{r}' - \mathbf{r}'')\right] = 1 \quad [3.27]$$

$$\langle \psi \mid \psi \rangle = \int d\mathbf{r} \int \frac{d\mathbf{k}'}{(2\pi)^3} \int \frac{d\mathbf{k}''}{(2\pi)^3} \langle \psi \mid \mathbf{k}' \rangle \langle \mathbf{k}'' \mid \psi \rangle \exp\left[+i\mathbf{r} \cdot (\mathbf{k}' - \mathbf{k}'')\right] = 1 \quad [3.28]$$

Let us make the following change in variables for the space coordinates:

$$\mathbf{s} = \mathbf{r}'' - \mathbf{r}', \quad \mathbf{r} = \frac{1}{2}(\mathbf{r}' + \mathbf{r}'') \quad [3.29]$$

and for the momentum:

$$\mathbf{q} = \mathbf{k}'' - \mathbf{k}', \quad \mathbf{k} = \frac{1}{2}(\mathbf{r}' + \mathbf{r}'') \quad [3.30]$$

Now, we can transform equations [3.27] and [3.28] into a simple normalization relationship[2]:

$$\langle \psi \mid \psi \rangle = \int \frac{d\mathbf{k}}{(2\pi)^3} \int d\mathbf{r} W(\mathbf{r}, \mathbf{p}, t) = 1 \quad [3.31]$$

2 Note that, as in Chapter 1, it is possible to normalize the Wigner distribution to the number of particles $N$ (see equation [1.20]).

where $W(\mathbf{r},\mathbf{p},t)$ is the Wigner distribution, which can now be represented in terms of state vectors as follows:

$$W(\mathbf{r},\mathbf{p},t) = \int \langle \mathbf{r}+\mathbf{s}/2 \mid \psi\rangle\langle\psi \mid \mathbf{r}-\mathbf{s}/2\rangle \exp(i\mathbf{k}\cdot\mathbf{s})d\mathbf{s} \qquad [3.32]$$

or, alternatively, in an equivalent manner:

$$W(\mathbf{r},\mathbf{p},t) = \int \langle \mathbf{k}+\mathbf{q}/2 \mid \psi\rangle\langle\psi \mid \mathbf{k}-\mathbf{q}/2\rangle \exp(-i\mathbf{q}\cdot\mathbf{r})\frac{d\mathbf{q}}{(2\pi)^3} \qquad [3.33]$$

Thus, the Wigner distribution is well normalized and indeed represents a distribution function in the phase space.

For illustrative purposes, Figure 3.3 shows some examples of Wigner distributions corresponding to a quantum harmonic oscillator for different values of energy levels. Note that it can take negative values, which makes it a quasi-distribution. It should also be noted that there are alternatives to the Wigner distribution. The *Husimi* distribution, for example, does not deal with the problem of negative values. However, it does not make it possible to find the current densities predicted by quantum mechanics. More information on the different representations of mechanics in phase space and a comparison between them can be found in Colomes et al. (2015).

### 3.3.2. *Weyl transformation*

Having defined a quantum distribution in the phase space, what remains is determining how to calculate the mean values of the observables from it. In quantum mechanics, observables are represented by operators and their mean values can be determined by knowing the wave vector $\mid \psi\rangle$. Let us consider, for example, a given observable, represented by an operator $\hat{A}$; the mean value of the operator in the state $\mid \psi\rangle$ is given by $\langle\hat{A}\rangle = \langle\psi|\hat{A}|\psi\rangle$. Now, a representation of quantum mechanics in the phase space requires calculation of the mean values of observables, which must be represented by functions of the variables of the phase space, namely the position and the momentum.

To find such a function, let us introduce two additional closure relationships of the form [3.20]. We can thus rewrite $\mid \psi\rangle \equiv \langle\hat{A}\rangle = \langle\psi|\hat{A}|\psi\rangle$ as follows:

$$\langle\hat{A}\rangle = \int \frac{d\mathbf{k}'}{(2\pi)^3}\int \frac{d\mathbf{k}''}{(2\pi)^3}\int d\mathbf{r}'\int d\mathbf{r}''\langle\psi \mid \mathbf{r}'\rangle\langle\mathbf{r}' \mid \mathbf{k}'\rangle\langle\mathbf{k}'|\hat{A}|\mathbf{k}''\rangle\langle\mathbf{k}'' \mid \mathbf{r}''\rangle\langle\mathbf{r}'' \mid \psi\rangle \qquad [3.34]$$

or, alternatively, in an equivalent manner, as follows:

$$\langle \hat{A} \rangle = \int \frac{d\mathbf{k}'}{(2\pi)^3} \int \frac{d\mathbf{k}^n}{(2\pi)^3} \int d\mathbf{r}' \int d\mathbf{r}'' \langle \psi \mid \mathbf{k}' \rangle \langle \mathbf{k}' \mid \mathbf{r}' \rangle \langle \mathbf{r}' | \hat{A} | \mathbf{r}'' \rangle \langle \mathbf{r}'' \mid \mathbf{k}'' \rangle \langle \mathbf{k}'' \mid \psi \rangle \tag{3.35}$$

Using the relationship [3.26], we can also write equation [3.34] as follows:

$$\langle \hat{A} \rangle = \int \frac{d\mathbf{k}'}{(2\pi)^3} \int \frac{d\mathbf{k}''}{(2\pi)^3} \int d\mathbf{r}' \int d\mathbf{r}'' \langle \psi \mid \mathbf{r}' \rangle \langle \mathbf{r}'' \mid \psi \rangle \langle \mathbf{k}' | \hat{A} | \mathbf{k}'' \rangle e^{-i\mathbf{r}' \cdot \mathbf{k} + i\hat{\mathbf{r}}'' \cdot \mathbf{k}''} \tag{3.36}$$

We can also rewrite equation [3.35] as follows:

$$\langle \hat{A} \rangle = \int \frac{d\mathbf{k}'}{(2\pi)^3} \int \frac{d\mathbf{k}''}{(2\pi)^3} \int d\mathbf{r}' \int d\mathbf{r}'' \langle \psi \mid \mathbf{k}' \rangle \langle \mathbf{k}'' \mid \psi \rangle \langle \mathbf{r}' | \hat{A} | \mathbf{r}'' \rangle e^{i\mathbf{r}'\mathbf{k} - i\mathbf{r}''\mathbf{k}''} \tag{3.37}$$

We can easily notice that from the change of variables [3.29] and [3.30], we have the relationship:

$$\exp\left(-i\mathbf{r}' \cdot \mathbf{k}' + i\mathbf{r}'' \cdot \mathbf{k}''\right) = \exp(i\mathbf{k} \cdot \mathbf{s} + i\mathbf{r} \cdot \mathbf{q}) \tag{3.38}$$

This makes it possible to rewrite, on the basis of equations [3.36] or [3.37], the mean value of the operator $\hat{A}$ as follows:

$$\langle \hat{A} \rangle = \int \frac{d\mathbf{k}}{(2\pi)^3} \int d\mathbf{r} a(\mathbf{r},\mathbf{k}) W(\mathbf{r},\mathbf{p},t) \tag{3.39}$$

where function $a(\mathbf{r},\mathbf{k})$ is given by:

$$a(\mathbf{r},\mathbf{k}) = \int \langle \mathbf{k} - \mathbf{q}/2 | \hat{A} | \mathbf{k} + \mathbf{q}/2 \rangle \exp(i\mathbf{q} \cdot \mathbf{r}) \frac{d\mathbf{q}}{(2\pi)^3} \tag{3.40}$$

It can also be rewritten as an integral in the position space as follows:

$$a(\mathbf{r},\mathbf{k}) = \int \langle \mathbf{r} - \mathbf{s}/2 | \hat{A} | \mathbf{r} + \mathbf{s}/2 \rangle \exp(-i\mathbf{k} \cdot \mathbf{s}) d\mathbf{s} \tag{3.41}$$

Thus, equation [3.39] expresses the mean value of an operator in the language of kinetic theory. Equation [3.39] has, in fact, a classic form: the mean value of an observable is expressed as the mean value of a function $a(\mathbf{r},\mathbf{k})$ of the phase space according to the Wigner distribution $W(\mathbf{r},\mathbf{r};t)$. The transformation allowing this (i.e. equations [3.40] or [3.41]) is called the *Weyl transform*. It may be interesting to note a trivial case: consider the case where the operator $\hat{A}$ in question is the identity operator. In this case, transformation [3.39] simply reduces to the normalization condition [3.31].

### 3.3.3. *Evolution of the Wigner distribution*

What remains to be done now is to determine an equation of evolution for $W(\mathbf{r},\mathbf{p},t)$, as we have done in the classic framework with the Boltzmann or Vlasov equation. For this purpose, we write two identical Schrödinger equations for the wave functions $\psi(\mathbf{r_1},t)$ and $\psi(\mathbf{r_2},t)$:

$$i\hbar\frac{\partial}{\partial t}\psi_j = H_j\psi_j \qquad [3.42]$$

where $j = 1,2$, the Hamiltonian being given by:

$$H_j = -\frac{\hbar^2}{2m}\frac{\partial^2}{\partial \mathbf{r}_j^2} + V(\mathbf{r}_j,t) \qquad [3.43]$$

By multiplying the equation for $j = 1$ by $\psi_2*$ (conjugate complex of $\psi_2$) and the conjugate complex of the equation for $j = 2$ by $\psi_1$, and by making the difference, we easily get:

$$i\hbar\frac{\partial}{\partial t}K = (H_1 - H_2)K \qquad [3.44]$$

where $K$ is defined by equation [3.18]. Now let us introduce the following new space variables:

$$\mathbf{r} = \frac{1}{2}(\mathbf{r}_1 + \mathbf{r}_2), \quad \mathbf{s} = \mathbf{r}_1 - \mathbf{r}_2$$
$$\Leftrightarrow \mathbf{r}_1 = (\mathbf{r} + \mathbf{s}/2), \quad \mathbf{r}_2 = (\mathbf{r} - \mathbf{s}/2) \qquad [3.45]$$

Thus:

$$\frac{\partial}{\partial \mathbf{r}_1} = \frac{1}{2}\frac{\partial}{\partial \mathbf{r}} + \frac{\partial}{\partial \mathbf{s}}, \quad \frac{\partial}{\partial \mathbf{r}_2} = \frac{1}{2}\frac{\partial}{\partial \mathbf{r}} - \frac{\partial}{\partial \mathbf{s}} \qquad [3.46]$$

After replacing in equation [3.44], we get:

$$i\hbar\frac{\partial}{\partial t}K = \left(-\frac{\hbar^2}{m}\frac{\partial}{\partial \mathbf{r}}\cdot\frac{\partial}{\partial \mathbf{s}} + \Delta V\right)K \qquad [3.47]$$

where we defined:

$$\Delta V \equiv (V_1 - V_2) = V(\mathbf{r} + \mathbf{s}/2, t) - V(\mathbf{r} - \mathbf{s}/2, t) \qquad [3.48]$$

Note that for a function $V(x+\Delta x)$, it is possible to develop a Taylor series around $x$, that is:

$$V(x+\Delta x) = V(x) + \sum_{n=1}^{\infty} \frac{(\Delta x)^n}{n!} \frac{\partial^n}{\partial x^n} V(x) \qquad [3.49]$$

Using the serial expansion of the exponential, we can also write[3]:

$$V(x+\Delta x) = \exp\left(\Delta x \frac{\partial}{\partial x}\right) V(x) \qquad [3.50]$$

Therefore, we have:

$$V_j \equiv V(\mathbf{r} \pm \mathbf{s}/2, t) = \exp\left(\pm\frac{\mathbf{s}}{2} \cdot \nabla\right) \qquad [3.51]$$

where the *plus* sign corresponds to $j=1$ and the *minus* sign to $j=2$. With this, $\Delta V$ can be rewritten in a more compact form as follows:

$$\Delta V = 2\sinh\left(\frac{1}{2}\mathbf{s} \cdot \nabla\right) \qquad [3.52]$$

where the hyperbolic sinus can be developed in series as follows:

$$\sinh \alpha = \sum_{l=0}^{\infty} \frac{1}{(2l+1)!} \alpha^{2l+1} \qquad [3.53]$$

This makes it possible to rewrite equation [3.47] as follows:

$$\left[i\hbar \frac{\partial}{\partial t} + \frac{\hbar^2}{m} \frac{\partial}{\partial \mathbf{r}} \cdot \frac{\partial}{\partial \mathbf{s}} - 2\sinh\left(\frac{1}{2}\mathbf{s} \cdot \nabla\right) V\right] K = 0 \qquad [3.54]$$

Equation [3.54] determines the evolution of the function $K$, whose Fourier transform represents the Wigner distribution (see equation [3.19]). By making a Fourier transform, we have the equivalent equation describing the evolution of the Wigner distribution:

$$i\hbar \left(\frac{\partial}{\partial t} + \frac{\hbar}{m}\mathbf{k} \cdot \frac{\partial}{\partial \mathbf{r}}\right) W = 2\sum_{l=0}^{\infty} \frac{W}{(2l+1)!} \left(\frac{\mathbf{s}}{2} \cdot \nabla\right)^{2l+1} V \qquad [3.55]$$

3 Let us recall that the exponential of an operator is understood in the sense of the development of the exponential as a series.

From the definition of the Wigner distribution [3.32], we can note that a derivation with respect to $\mathbf{k}$ is equivalent to a multiplication by $-is$, or again, more generally, the derivative $m$th of the Wigner distribution with respect to $\mathbf{k}$ is given by:

$$\frac{\partial^m W}{\partial \mathbf{k}^m} = (-i\mathbf{s})^m W \qquad [3.56]$$

Using this and the series expansion of the sine:

$$\sin x = \sum_{l=0}^{\infty} \frac{(-1)^l}{(2l+1)!} x^{2l+1} \qquad [3.57]$$

we finally obtain the equation describing the evolution of the Wigner distribution:

$$\left(\frac{\partial}{\partial t} + \frac{\hbar}{m}\mathbf{k} \cdot \frac{\partial}{\partial \mathbf{r}}\right) W = \frac{2}{\hbar} W (\sin \Lambda) V \qquad [3.58]$$

where:

$$\Lambda = \frac{1}{2} \frac{\overleftarrow{\partial}}{\partial \mathbf{k}} \cdot \frac{\overrightarrow{\partial}}{\partial \mathbf{r}} \qquad [3.59]$$

The arrows in equation [3.59] indicate the direction in which differential operators act. Thus, the derivatives with respect to $\mathbf{r}$ act on potential $V$ and the derivatives with respect to $\mathbf{k}$ (or, equivalently, with respect to $\mathbf{p}$) act on the Wigner distribution $W$.

Equation [3.58] represents the quantum generalization of the Vlasov equation. It is called the *Wigner–Moyal equation*. It can indeed be used in the same way and with (almost) the same ease as classic kinetic equations.

It may be interesting to study the classical limit of the Wigner–Moyal equation [3.58], which is none other than the Vlasov equation [1.25]. Indeed, taking the limit $\hbar \to 0$, equation [3.58] becomes:

$$\frac{\partial W}{\partial t} + \frac{\mathbf{p}}{m} \cdot \frac{\partial}{\partial \mathbf{r}} W - \frac{\partial}{\partial \mathbf{r}} V \cdot \frac{\partial}{\partial \mathbf{p}} W = 0 \qquad [3.60]$$

which is none other than the Vlasov equation (noting that $\mathbf{p} = m\mathbf{v}$ and that a conservative force derives from a potential, i.e. $\mathbf{F} = -\partial V / \partial \mathbf{r}$). More rigorously, the conventional limit amounts in fact to considering that potential $V$ varies very slowly in comparison with the de Broglie wavelength, which amounts to developing the right-hand side of equation [3.58] to the first order, that is:

$$sin(\Lambda) \approx \Lambda \qquad [3.61]$$

# 4

# General Relativity

*Everything is relative; and only that is absolute.*

Auguste Comte

In the previous chapters, we discussed the kinetic and hydrodynamic approaches in a classic framework and their generalization to the quantum mechanics framework. Another revolution in 20th century physics (in addition to quantum mechanics) is the theory of relativity, first special then general. Special relativity makes it possible to describe the behavior of particles whose speed approaches the speed of light in vacuum $c$. General relativity, on the other hand, makes it possible to unify special relativity with gravitation and describes the behavior of systems in the presence of intense gravitational fields.

Such extreme conditions of high-speed and intense gravitational fields are achieved in astrophysics. But, these systems involve large assemblies of particles that must be processed statistically. The kinetic and hydrodynamic approaches, extended to the context of relativity, represent essential tools in the study of such systems in relativistic regimes.

In this chapter, we will present relativistic formulations of the kinetic and hydrodynamic approaches. We will assume, throughout the chapter, that the reader has some degree of familiarity with the language of the theory of relativity. As a complement, we will present the central elements of the language of the theory of relativity in the Appendix (Minkowski space-time, coordinates, four-vectors, metrics, etc.).

## 4.1. Relativistic hydrodynamics

In this section, we will present the formulation of the hydrodynamic model within the context of the theory of relativity. We will start by presenting relativistic conservation equations. We will then develop formulations of the fluid model within the framework of special and then general relativity.

### 4.1.1. *Conservation equations*

Let us begin by writing the laws of conservation in the case of a relativistic fluid. There are five quantities that are conserved and the conservation laws are then written as:

$$(\rho u^{\mu})_{;\mu} = 0 \qquad [4.1]$$

and:

$$T^{\mu\nu}_{;\nu} = 0 \qquad [4.2]$$

where $\mu, \nu = 0, 1, 2, 3$. In the previous equations, $\nu$ represents the covariant derivative with respect to the $x^{\mu}$ coordinates (see Appendix). Note that the first equation [4.1] is scalar while the second [4.2] is vectorial with four components (four-vectorial). $\rho$ is the mass density (at rest) of the fluid, $u_{\mu}$ is the fluid velocity four-vector and $T^{\mu\nu}$ is the energy-momentum tensor. Note also that relationships [4.1] and [4.2] are in a covariant form, that is, they are true in any coordinate system. Equations [4.1] and [4.2] are formulated here in a very general framework and are valid both in the context of special relativity and in the context of general relativity. They are called *relativistic hydrodynamic equations*.

Here, we will only focus on perfect fluids, where the stress-energy tensor takes the following form:

$$T^{\mu\nu} = \rho h u^{\mu} u^{\nu} + p g^{\mu\nu} \qquad [4.3]$$

where $p$ represents the fluid pressure and $h$ represents the *specific enthalpy*, defined as:

$$h = \frac{(e+p)}{\rho} \qquad [4.4]$$

where $e$ is the energy density.

### 4.1.2. *Hydrodynamics in special relativity*

From the relativistic hydrodynamic equations [4.1] and [4.2], we can construct the fluid models of special and general relativity. In special relativity, the metric is that of Minkowski's space-time and is given by:

$$g^{\mu\nu} = \eta^{\mu\nu} = \mathrm{diag}(-1,1,1,1) \quad [4.5]$$

Note that in the case of a perfect fluid, the stress-energy tensor $T^{\mu\nu}$ can be interpreted as the flow of the $\mu$-pulse through a constant surface $x^{\nu}$, where component $\mu = 0$ represents the flow of the mean energy.

For a perfect fluid in Minkowski's space-time (the space-time of the theory of special relativity), we have:

$$\partial_{\mu}\left(\rho u^{\mu}\right) = 0 \quad [4.6]$$

and:

$$\partial_{\mu}\left(w u^{\mu} u^{\nu} + p\eta^{\mu\nu}\right) = 0 \quad [4.7]$$

These equations represent the hydrodynamic model within the special relativity framework. In equation [4.7], we have defined the relativistic enthalpy:

$$w = e + p = \rho h \quad [4.8]$$

As in the case of the non-relativistic hydrodynamic model (see Chapter 2), the relativistic model does not represent a closed system of equations and must be supplemented by an equation of state. The simplest case that will be discussed here is to write:

$$p = (\Gamma - 1)(e - \rho) \quad [4.9]$$

where $\Gamma$ is the *adiabatic index*, that is the ratio of the heat capacities at constant pressure and volume, that is:

$$c_P = \left(\frac{\partial h}{\partial T}\right)_p \quad c_V = \left(\frac{\partial(\rho/n)}{\partial T}\right)_v \quad [4.10]$$

In particular, we have (Del Zanna et al. 2003) $\Gamma = 5/3$ in the weakly relativistic case and $\Gamma = 4/3$ in the ultra-relativistic case.

It may be interesting to briefly discuss relativistic hydrodynamic equations in a more formal way. Note that in the fluid model, it is necessary to know the density, the fluid speed and the pressure. It may then be useful to define what are called *primitive variables* as follows:

$$\mathbf{V} = (\rho, v_j, p) \in \mathbb{R}^5 \qquad [4.11]$$

Note that it is just as possible to define these primitive variables by replacing pressure $p$ by energy $\varepsilon$, the two being linked by an equation of state.

With primitive variables [4.11], equations [4.1] and [4.2] no longer have a conservative form. Indeed, relativistic hydrodynamic equations can be written in a quasi-linear form as follows (Rezolla and Zanotti 2013):

$$\partial_t \mathbf{U} + \mathbf{A} \cdot \nabla \mathbf{U} + \mathbf{B} = 0 \qquad [4.12]$$

where $\mathbf{U} \in \mathbb{R}^N$ is the state vector and $\mathbf{A} \in \mathbb{R}^{N \times N}$ is diagonalizable with real eigenvalues and linearly independent eigenvectors. If we have $\mathbf{A} = \partial \mathbf{F} / \partial \mathbf{U}$, where $\mathbf{F}$ is a flux vector, the homogeneous solution of equation [4.12] is written as:

$$\partial_t \mathbf{U} + \nabla \cdot \mathbf{F}(\mathbf{U}) = 0 \qquad [4.13]$$

which has the form of a conservation equation. More generally, we can have a source term $\mathbf{S}$ and equation [4.13] then becomes:

$$\partial_t \mathbf{U} + \nabla \cdot \mathbf{F}(\mathbf{U}) = \mathbf{S} \qquad [4.14]$$

In the so-called *Valencia* formulation (Papadopoulos and Font 1998), the quantities retained are:

$$D = \rho W \qquad [4.15]$$

$$S_j = \rho h W^2 v_j \qquad [4.16]$$

$$E = \rho h W^2 - p \qquad [4.17]$$

and:

$$\tau = E - D \qquad [4.18]$$

where the Lorentz factor is given by:

$$W = \alpha u^t = \left(1 - v_i v^i\right)^{-1/2} \qquad [4.19]$$

Note that variables $E$ and $\tau$ both represent conserved energy. Only one of them should be used. The state vector is thus given by:

$$\mathbf{U}(\mathbf{V}) = (D, S_j, \tau) \in \mathbb{R}^5 \quad [4.20]$$

Projections of the stress-energy tensor will be denoted as $S^{\mu\nu}$ and are given by (Rezolla and Zanotti 2013):

$$S^{\mu\nu} = \rho h W^2 v^\mu v^\nu + p\gamma^{\mu\nu} \quad [4.21]$$

In Minkowski spacetime, equations [4.6] and [4.7] can be rewritten as equation [4.14] using the variable [4.20] for the flow:

$$\mathbf{F}^i(\mathbf{U}) = \left(\rho W v^i, \rho h W^2 v^i v^j + p\delta^{ij}, \rho h W^2 v^i - D v^i\right) \in \mathbb{R}^5 \quad [4.22]$$

Thus, Minkowski spacetime hydrodynamic equations have the form of conservation equations for $\mathbf{S} = 0$. A comparison with the non-relativistic equations (see Chapter 2) shows that the main difference is the introduction of the Lorentz factor $W$ (equation [4.19]).

### 4.1.3. *Hydrodynamic model in general relativity*

Here, we will briefly discuss the hydrodynamic model within the general relativity context. Since the mathematical language of general relativity is more sophisticated, we will not discuss the technical aspects. We will simply present the final equations that can be used in the hydrodynamic study of several systems, in the relativistic regime.

In the context of general relativity, fluid equations take the following general form (Rezolla and Zanotti 2013):

$$\partial_t(\sqrt{\gamma}\mathbf{U}) + \partial_i\left(\sqrt{\gamma}\mathbf{F}^i\right) = \mathbf{S} \quad [4.23]$$

which takes the form of equation [4.14], where variable $\mathbf{U}$ has been previously defined by relationship [4.20]. Flux vector $\mathbf{F}^i$ is given by:

$$\mathbf{F}^i = \begin{pmatrix} (\alpha v^i - \beta^i) D \\ \alpha S^i_j - \beta^i S_j \\ \alpha\left(S^i - D v^i\right) - \beta^i \tau \end{pmatrix} \quad [4.24]$$

As indicated by the notation, there are three flux vectors ($\mathbf{F}^1$, $\mathbf{F}^2$, $\mathbf{F}^3$), one for each direction in space. In the general case, source term $\mathbf{S}$ is not zero and is given by:

$$\mathbf{S} = \sqrt{\gamma}\begin{pmatrix} 0 \\ \frac{1}{2}\alpha S^{ik}\partial_j\gamma_{ik} + S_i\partial_j\beta^i - E\partial_j\alpha \\ \alpha S^{ij}K_{ij} - S^j\partial_j\alpha \end{pmatrix} \qquad [4.25]$$

Note that the source term remains valid even in the case of a time-dependent metric.

## 4.2. Relativistic kinetic theory

Here, we will discuss the kinetic approach within the context of the theory of relativity. After redefining the basic concepts of kinetic theory (distribution, phase space, etc.) in the relativistic context, we will follow the same steps presented in Chapter 1 in the non-relativistic case. The approach discussed here is similar to that of Vereshchagin and Aksenov (2017) where more details on relativistic kinetic methods and their application in astrophysics and cosmology can be found.

### 4.2.1. *Basic concepts*

To build a relativistic kinetic theory, the basic ingredient is the definition of a distribution function in phase space. How can this be done?

There are actually two methods. The first consists of considering a phase space composed of position four-vectors $x^\mu = (ct, \mathbf{x})$ and stress four-vectors $p^\mu = (p^0, \mathbf{p})$, considered to be independent variables. Thus, unlike the non-relativistic phase space composed of six dimensions (for a three-dimensional system), this one is an eight-dimensional space $\mathscr{M}^8$. A distribution function $\mathscr{F}(x,p)$ can thus be defined in this space, so that:

$$j^\mu(x) \equiv c\int \mathscr{F}(x,p)p^\mu d^4p \qquad [4.26]$$

where $j^\mu(x)$ is the current four-vector. In what follows, to simplify notations, $x^\mu$ and $p^\mu$ will be denoted as $x$ and $p$, respectively. Such a definition of the distribution function is indeed invariant under the Lorentz transformation. However, it does not have the meaning of density of particles in a region of space (unlike the non-relativistic distribution function), because it is defined in the eight-dimensional phase space $\mathscr{M}^8$.

Another alternative consists of defining a relativistic distribution function $f(\mathbf{p},\mathbf{x},t)$ in the six-dimensional phase space. As in the non-relativistic case (see Chapter 1), such a distribution depends on seven variables, namely position, momentum or stress and time. As in the non-relativistic case, such a distribution represents a density of particles in a given region, and satisfies the relationship:

$$N \equiv \int_{\mathscr{M}^6} f(\mathbf{p},\mathbf{x},t) d^3p d^3x \qquad [4.27]$$

we can see that, as in the non-relativistic case (see equation [1.20] of Chapter 1), quantity $f(x,p)d^3pd^3x$ represents the average number of particles located in the region $(\mathbf{x},\mathbf{x}+d^3x)$ and with a momentum in domain $(\mathbf{p},\mathbf{p}+d^3p)$ at the instant $t$.

Similarly, by integrating the distribution function $f(\mathbf{p},\mathbf{x},t)$ on the momentum, the particle density is obtained, as in the non-relativistic case:

$$n(\mathbf{x},t) = \int_{-\infty}^{+\infty} f d^3p \qquad [4.28]$$

Thus, if $\mathbf{x}_i(t)$ and $\mathbf{p}_i(t)$ represent, respectively, the position and the momentum of a particle $i$, the distribution $f(\mathbf{p},\mathbf{x},t)$ represents an average particle density, that is:

$$f(\mathbf{p},\mathbf{x},t) = \left\langle \sum_i \delta^{(3)}\left[\mathbf{p}-\mathbf{p}_i(t)\right] \delta^{(3)}\left[\mathbf{x}-\mathbf{x}_i(t)\right] \right\rangle \qquad [4.29]$$

Thus, it appears that, although it is natural to work in an eight-dimensional phase space $\mathscr{M}^8$, results can be represented in subspace $\mathscr{M}^6$, defined by $p^\mu p_\mu = m^2c^2$ with $p^0 > 0$. It is thus possible to introduce the function:

$$\mathscr{F}(x,p) = 2\Theta\left(p^0\right)\delta\left(p^\mu p_\mu - m^2c^2\right) f(\mathbf{px},t) \qquad [4.30]$$

where $\Theta\left(p^0\right)$ is the Heaviside function[1]. By using the relationship:

$$\delta[Z(x)] = \sum_i \left|\frac{dZ}{dx}\right|^{-1} \delta\left(x-x_i\right) \qquad [4.32]$$

1 The Heaviside function is discontinuous at 0 and takes the value 1 for all positive real numbers and the value 0 for strictly negative real numbers, that is,

$$\Theta(x) = \begin{cases} 0 & \text{if } x<0 \\ \text{not defined} & \text{if } x=0 \\ 1 & \text{if } x>0 \end{cases} \qquad [4.31]$$

where $x_i$ represents the roots of $Z(x) = 0$. We can rewrite the relationship [4.30] using the relationship [4.29] as follows:

$$\mathscr{F}(x,p) = \left\langle \sum_i \frac{1}{p_i^0(t)} \delta^3 \left[\mathbf{p} - \mathbf{p}_i(t)\right] \delta \left[p - p_i^0(t)\right] \delta^3 \left[\mathbf{x} - \mathbf{x}_i(t)\right] \right\rangle \quad [4.33]$$

By introducing a Dirac delta and integrating over time, we have:

$$\mathscr{F}(x,p) = \int dt \left\langle \sum_i \frac{1}{p_i^0(t_i)} \delta\left[t - t_i\right] \delta^3 \left[\mathbf{p} - \mathbf{p}_i(t_i)\right] \right. \\ \left. \times \delta \left[p - p_i^0(t_i)\right] \delta^3 \left[\mathbf{x} - \mathbf{x}_i(t_i)\right] \right\rangle \quad [4.34]$$

Using the relationship between time $t_i$ and eigentime $d\tau_i = mcdt_i/p_i^0(t_i)$, we can express $\mathscr{F}(x,p)$ as follows:

$$\mathscr{F}(x,p) = \frac{1}{mc} \int d\tau \left\langle \sum_i \delta^4 \left[x - x_i(\tau)\right] \delta^4 \left[p - p_i(\tau)\right] \right\rangle \quad [4.35]$$

where $x_i(\tau)$ and $p_i(\tau)$ are trajectories in the eight-dimensional phase space $\mathscr{M}^8$.

As in the non-relativistic case, we can observe that macroscopic quantities can depend on position and time, but not on momentum (or velocity). Thus, for a microscopic quantity $\chi(x,p)$ defined in the phase space $\mathscr{M}^6$, it is possible to introduce a macroscopic quantity $A(x)$ by integrating on the momentum as follows:

$$A(x) \equiv \int_{-\infty}^{+\infty} \chi(x,p) d^3p \quad [4.36]$$

It should also be noted that, from the relationship [4.26], an invariant quantity can be defined as follows:

$$j^\mu(\mathbf{x},t) \equiv c \int p^\mu f \frac{d^3p}{p^0} \quad [4.37]$$

This is the first stage of distribution. Its spatial part represents the flow:

$$\mathbf{j}(\mathbf{x},t) = c \int \mathbf{v} f d^3p \quad [4.38]$$

where $\mathbf{v} = c\mathbf{p}/p^0$ (the velocity of a relativistic particle of momentum $\mathbf{p}$). The moment of order 2 may also be established in an equivalent manner:

$$T^{\mu\nu}(\mathbf{x},t) \equiv c\int p^{\mu}p^{\nu}f\frac{d^3p}{p^0} \qquad [4.39]$$

where $T^{\mu\nu}$ is the stress-energy tensor. Similarly, it may be useful to define the entropy flux as follows:

$$S^{\mu}(\mathbf{x},t) \equiv -k_B c\int p^{\mu}f\frac{d^3p}{p^0}\left[\log\left(h^3 f\right) - 1\right] \qquad [4.40]$$

where quantity $h$ is introduced to leave the logarithm argument dimensionless.

It should be noted, however, that, unlike the non-relativistic case, macroscopic velocity can be defined in several different ways. The most useful are (i) the definition by Eckart (1940):

$$U_{\mathrm{E}}^{\mu} \equiv \frac{c j^{\mu}}{\sqrt{j^{\mu} j_{\mu}}} \qquad [4.41]$$

and (ii) that by Landau and Lifshitz (1959):

$$U_{\mathrm{LL}}^{\mu} \equiv \frac{cT^{\mu\nu}U_{\nu}}{\sqrt{U_{\rho}T^{\rho\sigma}T_{\sigma\tau}U^{\tau}}} \qquad [4.42]$$

The first definition represents the mean velocity of particles, while the second represents the mean velocity of the stress-energy transfer.

### 4.2.2. *Formulation of kinetic theory*

Let us consider, for simplicity, the case of free particles (without interactions). We can define the following quantity:

$$\Delta J = \frac{1}{c}\int_{\Delta^3\sigma} d^3\sigma_{\mu}\, j^{\mu} = \int_{\Delta^3\sigma} d^3\sigma_{\mu}\int\frac{d^3p}{p^0}p^{\mu}f \qquad [4.43]$$

where $d^3\sigma_{\mu}$ is an oriented surface element and $\Delta^3\sigma$ is a small surface element. Thus, the last equality in relationship [4.43] follows from relationship [4.37].

In the Lorentz coordinate system, quantity $d^3\sigma_\mu$ has coordinates $(d^3x,0,0,0)$. In this coordinate system, we have:

$$\Delta J = \int_{\Delta^3\sigma} \int f(x,p) d^3p d^3x \qquad [4.44]$$

This quantity represents the mean number of universe lines crossing $\Delta^3\sigma$. Considering the universe lines with a momentum in the domain $\Delta^3 p$ around $\mathbf{p}$, we have:

$$\Delta J = \int_{\Delta^3\sigma} \int_{\Delta^3 p} f(x,p) d^3p d^3x \qquad [4.45]$$

Let us consider the universe lines given by the relationship [4.43], which later cross an element $\Delta^3\hat{\sigma}$. Since we assume that there are no collisions, we can write:

$$\int_{\Delta^3\hat{\sigma}} d^3\sigma_\mu \int_{\Delta^3 p} \frac{d^3p}{p^0} p^\mu f - \int_{\Delta^3\sigma} d^3\sigma_\mu \int_{\Delta^3 p} \frac{d^3p}{p^0} p^\mu f = 0 \qquad [4.46]$$

or alternatively, in an equivalent manner:

$$\int_{\Delta^3 x} d^3\sigma_\mu \int_{\Delta^3 p} \frac{d^3p}{p^0} p^\mu f = 0 \qquad [4.47]$$

where $\Delta^3 x$ is the area of the Minkowski space element $\Delta^4 x$. Applying Gauss's law, we have:

$$\int_{\Delta^4 x} d^4x \int_{\Delta^3 p} \frac{d^3p}{p^0} p^\mu \partial_\mu f = 0 \qquad [4.48]$$

where:

$$\partial_\mu = (c^{-1}\partial/\partial t, \nabla) \qquad [4.49]$$

and $\Delta^3 x$ and $\Delta^3 p$ are any hypersurfaces of the phase space.

We are looking for a kinetic equation, that is, an equation governing the temporal variation of the distribution function. In the absence of interactions, the latter simply expresses the conservation of the four-vector $p^\mu f$.

It then follows from relationship [4.48] and is given by:

$$p^\mu \partial_\mu f = 0 \qquad [4.50]$$

This can be demonstrated more transparently by rewriting it in vector form (and dividing by $p^0/c$).

Then, we have:

$$\frac{df}{dt} \equiv \frac{\partial f}{\partial t} + \mathbf{v} \cdot \nabla f = 0 \qquad [4.51]$$

Remember that this equation is only valid in the absence of collisions and interactions.

More generally, we have:

$$p^{\mu} \partial_{\mu} f + mF^{\mu} \frac{\partial f}{\partial p^{\mu}} = \mathrm{St}[f] \qquad [4.52]$$

where $F^{\mu}$ represents the four-vector force and $\mathrm{St}[f]$ represents the collision term which will be discussed in section 4.2.3.

### 4.2.3. *Collision integral*

Consider the collision between two particles, of mass $m_1$ and $m_2$, and with momentum $p_1^{\mu}$ and $p_2^{\mu}$ before collisions and $p_1'^{\mu}$ and $p_2'^{\mu}$, after collisions. The stress-energy four-vector is conserved and, therefore, we have:

$$p_1^{\mu} + p_2^{\mu} = p_1'^{\mu} + p_2'^{\mu} \qquad [4.53]$$

It should now be noted that the number of collisions of this type is proportional to the number of particles per unit volume of momentum $p_1^{\mu}$ ($p_2^{\mu}$) in a domain $d^3p_1$ ($d^3p_2$). It is also proportional to the intervals $d^3p_1'^{\mu}$ and $d^3p_2'^{\mu}$ and $d^4x$. The coefficients of proportionality depend solely on stress-energy four-vectors before and after collision, and will be denoted as $W(p_1, p_2 \mid p_1', p_2') / (p_1^0 p_2^0 p_1'^0 p_2'^0)$. As in the non-relativistic case, the quantity $W(p_1, p_2 \mid p_1', p_2')$ represents a *transition rate*: it takes the transitions of particles leaving volume $d^3p_1$ around $p_1^{\mu}$ into account. Similarly, quantity $W(p_1', p_2' \mid p_1, p_2)$ takes particles not located in volume $d^3p_1$ around $p_1^{\mu}$, before collision, and which are there after collision into account. Thus, the Boltzmann equation in the presence of collisions is written as:

$$\begin{aligned}\int_{\mathscr{V}}\int_{\mathscr{P}} p^{\mu}\partial_{\mu} f \frac{d^3p_1}{p_1^0} d^4x = \frac{1}{2}\int_{\mathscr{V}}\int_{\mathscr{P}}\int \frac{d^3p_1}{p_1^0}\frac{d^3p_1'}{p_1'^0}\frac{d^3p_2}{p_2^0}\frac{d^3p_2'}{p_2'^0} \\ \times \left[f(x,p_1') f(x,p_2') W(p_1',p_2' \mid p_1,p_2)\right. \\ \left. - f(x,p_1) f(x,p_2) W(p_1,p_2 \mid p_1',p_2')\right] d^4x\end{aligned} \qquad [4.54]$$

where $\mathscr{V}$ and $\mathscr{P}$ are volumes in the space of coordinates and momentum, respectively. Equivalently, we can also write:

$$\begin{aligned} p^\mu \partial_\mu f &= \frac{1}{2} \int \frac{d^3 p'_1}{p_1^0} \frac{d^3 p_2}{p_2^0} \frac{d^3 p'_2}{p_2'^0} \\ &\times \left[f\left(x, p'_1\right) f\left(x, p'_2\right) W\left(p'_1, p'_2 \mid p_1, p_2\right) \right. \\ &\left. - f\left(x, p_1\right) f\left(x, p_2\right) W\left(p_1, p_2 \mid p'_1, p'_2\right)\right] d^4 x \end{aligned} \qquad [4.55]$$

So, in vector form, we have:

$$\begin{aligned} \frac{\partial f}{\partial t} + \mathbf{v} \cdot \nabla f &= \frac{1}{2} \int d^3 p'_1 d^3 p_2 d^3 p'_2 \left[f\left(x, p'_1\right) f\left(x, p'_2\right) w_{p'_1 p'_2; p_1 p_2} \right. \\ &\left. - f\left(x, p_1\right) f\left(x, p_2\right) w_{p_1 p_2; p'_1 p'_2}\right] \end{aligned} \qquad [4.56]$$

where:

$$w_{p_1 p_2; p'_1 p'_2} = cW\left(p_1, p_2 \mid p'_1, p'_2\right) / \left(p_1^0 p_2^0 p_1'^0 p_2'^0\right) \qquad [4.57]$$

Equation [4.56] is the *relativistic Boltzmann equation* in the presence of collisions (it thus generalizes equation [4.50]). Note that factor $1/2$ in the right-hand side of the equation is due to the indistinguishability of particles.

Before concluding this section, note that in the case of particles obeying a quantum statistic (Fermi-Dirac or Bose-Einstein), the collision term can be generalized as follows (Vereshchagin and Aksenov 2017):

$$\begin{aligned} \mathrm{St}[f] &= \frac{1}{2} \int \frac{d^3 p'_1}{p_1'^0} \frac{d^3 p_2}{p_2^0} \frac{d^3 p'_2}{p_2^0} \\ &\times \left\{f\left(x, p'_1\right) f\left(x, p'_2\right) \left[1 + \theta \bar{n}\left(x, p_1\right)\right] \left[1 + \theta \bar{n}\left(x, p_2\right)\right] W\left(p'_1, p'_2 \mid p_1, p_2\right) \right. \\ &\left. - f\left(x, p_1\right) f\left(x, p_2\right) \left[1 + \theta \bar{n}\left(x, p'_1\right)\right] \left[1 + \theta \bar{n}\left(x, p'_2\right)\right] W\left(p_1, p_2 \mid p'_1, p'_2\right)\right\} \end{aligned} \qquad [4.58]$$

where:

$$f(x, p) = g\bar{n}(x, p) / (2\pi\hbar)^3 \qquad [4.59]$$

and $g$ is the degeneration factor and quantity $\bar{n}(x, p)$ is the *degeneracy factor*. Cases $\theta = +1$, $\theta = -1$, and $\theta = 0$ correspond to the Bose–Einstein, Fermi–Dirac and Boltzmann statistics, respectively.

### 4.2.4. Boltzmann equation in general relativity

It may be useful to generalize the Boltzmann equation in the context of general relativity. To do this, let us consider the eight-dimensional phase space $\mathscr{M}^8$ and define $\mathscr{F}^K(x,p)$ the Klimontovich function:

$$\mathscr{F}^K(x,p) = \frac{1}{mc}\sum_i \int d\tau \delta^4 \left[x - x_i(\tau)\right] \delta^4 \left[p - p_i(\tau)\right] \quad [4.60]$$

whose mean value is given by the distribution function [4.35], namely:

$$\mathscr{F}(x,p) = \left\langle \mathscr{F}^K(x,p) \right\rangle \quad [4.61]$$

In general relativity, equations of momentum in a gravitational field are given by:

$$mc\frac{dx^\mu}{ds} = p^\mu \quad mc\frac{dp^\mu}{ds} = -\Gamma^\mu_{\nu\lambda} p^\nu p^\lambda \quad [4.62]$$

where:

$$ds = \left(g_{\mu\nu} dx^\mu dx^\nu\right)^{1/2} = c d\tau \quad [4.63]$$

is the space-time interval element, $g_{\mu\nu}$ is the metric tensor and $\Gamma^\mu_{\nu\lambda}$ are the Christoffel symbols (see the Appendix). Using the following property of the Dirac delta function:

$$\frac{d}{ds}\delta[x - g(s)] = -\frac{d}{dx}\delta[x - g(s)]\frac{dg}{ds} \quad [4.64]$$

as well as:

$$\int ds \frac{d}{ds}\left\{\delta^4 \left[x - x_i(s)\right] \delta^4 \left[p - p_i(s)\right]\right\} = 0 \quad [4.65]$$

we get:

$$\frac{\partial\left(p^\mu \mathscr{F}^K\right)}{\partial x^\mu} - \frac{\partial}{\partial p^\mu}\left(\Gamma^\mu_{\nu\lambda} p^\nu p^\lambda \mathscr{F}^K\right) = 0 \quad [4.66]$$

Also, we can show the following identity (Vereshchagin and Aksenov 2017):

$$\frac{\partial p^\mu}{\partial x^\mu} - \frac{\partial}{\partial p^\mu}\left(\Gamma^\mu_{\nu\lambda} p^\nu p^\lambda\right) = 0 \quad [4.67]$$

Applying the latter to relationship [4.66], we have:

$$p^{\mu}\frac{\partial \mathscr{F}}{\partial x^{\mu}} - \Gamma^{\mu}_{\nu\lambda}p^{\nu}p^{\lambda}\frac{\partial \mathscr{F}}{\partial p^{\mu}} = 0 \qquad [4.68]$$

This is the Boltzmann equation without collisions (Vlasov equation) in the relativistic context, concerning distribution $\mathscr{F}$ defined in the eight-dimensional phase space $\mathscr{M}^8$.

In the same way, it is possible to establish a kinetic equation governing the evolution of distribution $f$ defined in the six-dimensional phase space $\mathscr{M}^6$. This can be obtained by using relationship [4.30], making it possible to go from $\mathscr{F}$ to $f$ and vice versa. It is then sufficient to integrate equation [4.68] on $p^0$ to get:

$$p^{\mu}\frac{\partial f}{\partial x^{\mu}} - \Gamma^{i}_{\nu\lambda}p^{\nu}p^{\lambda}\frac{\partial f}{\partial p^{i}} = 0 \qquad [4.69]$$

The latter, like equation [4.68], are valid in the absence of collisions. In the presence of collisions, the right-hand side is no longer zero, but takes into account a change in the distribution function due to collisions. Thus, it will be possible to generally write:

$$p^{\mu}\frac{\partial f}{\partial x^{\mu}} - \Gamma^{i}_{\nu\lambda}p^{\nu}p^{\lambda}\frac{\partial f}{\partial p^{i}} = \mathrm{St}[f] \qquad [4.70]$$

It may be interesting to compare the Boltzmann equation in the context of general relativity [4.70] with equation [4.52], formulated in the context of special relativity. The curved nature of space-time in general relativity appears to give rise to a term similar to the force term in equation [4.52].

Note that, as in the non-relativistic case, many approximations of the collision term exist and can be of great use because they make it possible to obtain approximate analytical solutions. In particular, the BGK approximation (see section 1.3.2 in Chapter 1) can be extended to the relativistic context by writing the collision term as follows:

$$\mathrm{St}[f] = -p^{\mu}U_{\mu}\frac{1}{\tau}\left[f(x,p,t) - f_{eq}(x,p)\right] \qquad [4.71]$$

where $f_{eq}(x,p)$ is the equilibrium distribution and $\tau$ the relaxation time.

### 4.2.5. *H-theorem*

As in the non-relativistic case, the H-theorem can be formulated in the relativistic context. As a reminder, in the non-relativistic case, the H-theorem stipulated that there is a quantity (entropy) that cannot decrease during the evolution of a system toward thermal equilibrium. In relativity, the quantity that cannot decrease is the divergence of the entropy flux $S^\mu$ defined by relationship [4.40], that is:

$$\sigma(x) \equiv \partial_\mu S^\mu \qquad [4.72]$$

From the definition of $S^\mu$ [4.40], relationship [4.72] is written as:

$$\sigma = -k_{BC} \int \frac{d^3 p}{p^0} \left[\log\left(h^3 f\right)\right] p^\mu \partial_\mu f \qquad [4.73]$$

Using the relativistic Boltzmann equation [4.52], we then have:

$$\sigma = -k_{BC} \int \frac{d^3 p}{p^0} \left[\log\left(h^3 f\right)\right] \mathrm{St}[f] + k_{BC} \int \frac{d^3 p}{p^0} \left[\log\left(h^3 f\right)\right] F^\mu \frac{\partial f}{\partial p^\mu} \qquad [4.74]$$

We will assume that the force satisfies the following relationships:

$$p^\mu F_\mu = 0, \qquad \frac{\partial F^\mu}{\partial p^\mu} = 0 \qquad [4.75]$$

These conditions express the fact that force is mechanical and does not affect the resting mass of the particle. Thus, the second integral in relationship [4.74] can be written as:

$$2k_{BC} \int d^4 p \frac{\partial}{\partial p^\mu} \left\{\Theta\left(p^0\right) \delta\left(p^\mu p_\mu - m^2 c^2\right) f \left[\log\left(h^3 f\right) - 1\right] F^\mu\right\} \qquad [4.76]$$

and is canceled out if the integrator decreases sufficiently rapidly for large values of the momentum. Similarly, the first integral of relationship [4.74] can be written as:

$$\sigma = -\frac{1}{4} k_{BC} \sum_{i,j,k,l} \int \frac{d^3 p_i}{p_i^0} \frac{d^3 p_j}{p_j^0} \frac{d^3 p_k}{p_k^0} \frac{d^3 p_l}{p_l^0} \left[\log\left(\frac{f_k f_l}{f_i f_j}\right)\right] f_i f_j W_{ij|kl} \qquad [4.77]$$

We can use the following identity (Vereshchagin and Aksenov 2017):

$$\sum_{i,j,k,l} \int \frac{d^3 p_i}{p_i^0} \frac{d^3 p_j}{p_j^0} \frac{d^3 p_k}{p_k^0} \frac{d^3 p_l}{p_l^0} \left(f_k f_l - f_i f_j\right) W_{ij|kl} = 0 \qquad [4.78]$$

to get:

$$\sigma = \frac{1}{4} k_B C \sum_{i,j,k,l} \int \frac{d^3 p_i}{p_i^0} \frac{d^3 p_j}{p_j^0} \frac{d^3 p_k}{p_k^0} \frac{d^3 p_l}{p_l^0} A f_i f_j W_{ij|kl} \qquad [4.79]$$

where:

$$A = \frac{f_k f_l}{f_i f_j} - \log\left(\frac{f_k f_l}{f_i f_j}\right) - 1 \qquad [4.80]$$

It can be observed that this quantity is non-negative, $A \geq 0$. Relationship [4.79] then implies that:

$$\sigma \geq 0 \qquad [4.81]$$

This is the *relativistic H-theorem*!

It can also be noted that $\sigma = 0$ is true only for:

$$f_i(x, p_i) f_j(x, p_j) = f_k(x, p_k) f_l(x, p_l) \qquad [4.82]$$

This is the equilibrium condition. In the context of relativity, however, equilibrium distribution is no longer given by the Maxwell–Boltzmann distribution, but by a generalization of it: the Jüttner distribution, whose properties will be studied in section 4.2.6.

### 4.2.6. *Equilibrium state: Jüttner distribution*

The equilibrium state is reached when relationship [4.82] is satisfied. It can easily be seen from the relativistic Boltzmann equation [4.52] that this happens when the right-hand side is zero. A similar reasoning to the non-relativistic case can be followed to establish the equilibrium distribution in the relativistic context.

Let us start by rewriting relationship [4.82] in a logarithmic form as follows:

$$\log\left(h^3 f_1\right) + \log\left(h^3 f_2\right) = \log\left(h^3 f_1'\right) + \log\left(h^3 f_2'\right) \qquad [4.83]$$

As in the non-relativistic case, this means that quantity $\log\left(h^3 f\right)$ is an invariant, and can thus be written as a linear combination of invariants, which are, here, a constant and $p^\mu$. The equilibrium distribution is therefore in the form:

$$f^{eq} = \frac{1}{h^3} \exp\left[a(x) + b_\mu(x) p^\mu\right] \qquad [4.84]$$

The coefficients of distribution can be found by observing that if the right-hand side of the relativistic Boltzmann equation [4.52] cancels itself out, the left-hand side must also be zero for $f^{eq}$. Thus, we have:

$$p^{\mu}\partial_{\mu}a(x)+p^{\mu}p^{\nu}\partial_{\mu}b_{\nu}(x)+mb_{\mu}(x)F^{\mu}(x,p)=0 \qquad [4.85]$$

which must be satisfied for any $p^{\mu}$. Considering the absence of an external field, the distribution is given in the explicit form:

$$f^{eq}(p)=\frac{1}{h^3}\exp\left[\frac{\phi-p^{\mu}U_{\mu}}{k_BT}\right] \qquad [4.86]$$

where $\phi$, $T$ and $\mu$ are parameters to be determined. Distribution [4.86] is called *Jüttner distribution*.

From the Jüttner distribution, it may be interesting to establish certain macroscopic magnitudes, describing a relativistic system at the equilibrium state (and in the absence of an external field). First, using the definition [4.37] as well as:

$$j^{\mu}=nU^{\mu},\quad U^{\mu}U_{\mu}=c^2 \qquad [4.87]$$

density can be written as follows:

$$n=\frac{j^{\mu}U_{\mu}}{c^2}=\frac{1}{ch^3}\exp\left(\frac{\phi}{k_BT}\right)\int\frac{d^3p}{p^0}p^{\mu}U_{\mu}\exp\left(\frac{-p^{\nu}U_{\nu}}{k_BT}\right) \qquad [4.88]$$

Since it is scalar, the integral can be evaluated in the Lorentz coordinate system, where $U^{\mu}=(c,0,0,0)$. By introducing the following variables:

$$\theta=k_BT/\left(mc^2\right),\quad \nu=\phi/\left(mc^2\right),\quad y=c\sqrt{\mathbf{p}^2+m^2c^2}/\left(k_BT\right) \qquad [4.89]$$

the equilibrium density is obtained as follows:

$$n^{eq}=\frac{1}{2\pi^2\lambda_C^3}\exp\left(\frac{\nu}{\theta}\right)\theta K_2\left(\theta^{-1}\right) \qquad [4.90]$$

where:

$$K_n\left(\theta^{-1}\right)=\frac{2^{n-1}(n-1)!}{(2n-2)!}z^{-n}\int_z^{\infty}dy\left(y^2-\theta^{-2}\right)^{n-\frac{3}{2}}y\exp(-y) \qquad [4.91]$$

is the *modified Bessel function of the second kind* and $\lambda_C\equiv h/mc$.

Similarly, using the definition of the stress-energy tensor:

$$T^{\mu\nu} = c^{-2}\rho U^{\mu}U^{\nu} - p\Delta^{\mu\nu} \quad [4.92]$$

where $\Delta^{\mu\nu} = g^{\mu\nu} - c^{-2}U^{\mu}U^{\nu}$ is the projection operator, we can establish the energy density:

$$\rho = \frac{T^{\mu\nu}U_{\mu}U_{\nu}}{c^2} = \frac{1}{c}\int \frac{d^3p}{p^0}\left(p^{\mu}U_{\mu}\right)^2 f^{eq} \quad [4.93]$$

or:

$$\rho = \frac{mc^2}{2\pi^2\lambda_C^3}\exp\left(\frac{v}{\theta}\right)\left[3\theta^2K_2\left(\theta^{-1}\right) + \theta K_1\left(\theta^{-1}\right)\right] \quad [4.94]$$

We can also find pressure $P$ in an equivalent manner:

$$P = -\frac{1}{3}T^{\mu\nu}\Delta_{\mu\nu} = -\frac{c}{3}\int \frac{d^3p}{p^0}p^{\mu}p^{\nu}\Delta_{\mu\nu}f^{eq} \quad [4.95]$$

or:

$$P = \frac{mc^2}{2\pi^2\lambda_C^3}\exp\left(\frac{v}{\theta}\right)\theta^2K_2\left(\theta^{-1}\right) \quad [4.96]$$

It may be useful to define *enthalpy* $h_e$ as follows:

$$h_e = \frac{\rho + P}{n} = mc^2\frac{K_3\left(\theta^{-1}\right)}{K_2\left(\theta^{-1}\right)} \quad [4.97]$$

Entropy density $s$ can be written as:

$$s = \frac{S^{\mu}U_{\mu}}{c^2} = -\frac{k_B}{c}\exp\left(\frac{\phi}{k_BT}\right)\int \frac{d^3p}{p^0}p^{\mu}U_{\mu}\left(\frac{\varphi - p^{\nu}U_{\nu}}{k_BT} - 1\right)\exp\left(\frac{-p^{\nu}U_{\nu}}{k_BT}\right) \quad [4.98]$$

Or, using equilibrium density [4.90] and energy density [4.94], we have:

$$s = \frac{1}{T}(\rho - \phi n) + k_Bn \quad [4.99]$$

Finally, let us define the *adiabatic index* $\Gamma \equiv c_p/c_v$, as the ratio of heat capacities at constant pressure and volume, that is:

$$c_P = \left(\frac{\partial h_e}{\partial T}\right)_p \quad c_V = \left(\frac{\partial (\rho/n)}{\partial T}\right)_v \quad [4.100]$$

so, we have:

$$\frac{\Gamma}{\Gamma - 1} = \theta^{-2} + 5\left(\frac{h_e}{mc^2\theta}\right) - \left(\frac{h_e}{mc^2\theta}\right)^2 \quad [4.101]$$

Combining relationships [4.90], [4.94], [4.96], [4.97] and [4.99], we have:

$$\begin{aligned} P &= nk_B T \\ P &= (\Gamma - 1)\rho \\ \phi &= h_e - Ts \end{aligned} \quad [4.102]$$

which are none other than the laws of an ideal gas by denoting $T$ as the temperature and $\phi$ as the chemical potential.

It is useful to consider the Jüttner distribution [4.86] in the local coordinate system at rest. We can then express this as:

$$f_0^{eq} = \frac{1}{h^3}\exp\left(-\frac{\gamma}{\theta}\right) \quad [4.103]$$

where $\gamma \equiv p^0/mc$ is the Lorentz factor. By comparing equilibrium density [4.90] with relationship [4.28], we can write the distribution function in terms of the Lorentz factor:

$$f_0^{eq}(\gamma) = \frac{1}{\theta K_2(\theta^{-1})}\gamma\sqrt{\gamma^2 - 1}\exp\left(-\frac{\gamma}{\theta}\right) \quad [4.104]$$

or, in terms of $\beta = |\mathbf{v}|/c$, as follows:

$$f_0^{eq}(\beta) = \frac{1}{\theta K_2(\theta^{-1})}\gamma^5\beta^2\exp\left(-\frac{1}{\sqrt{1-\beta^2}\theta}\right) \quad [4.105]$$

The latter is represented in Figure 4.1, for different values of $\theta = k_B T/\left(mc^2\right)$.

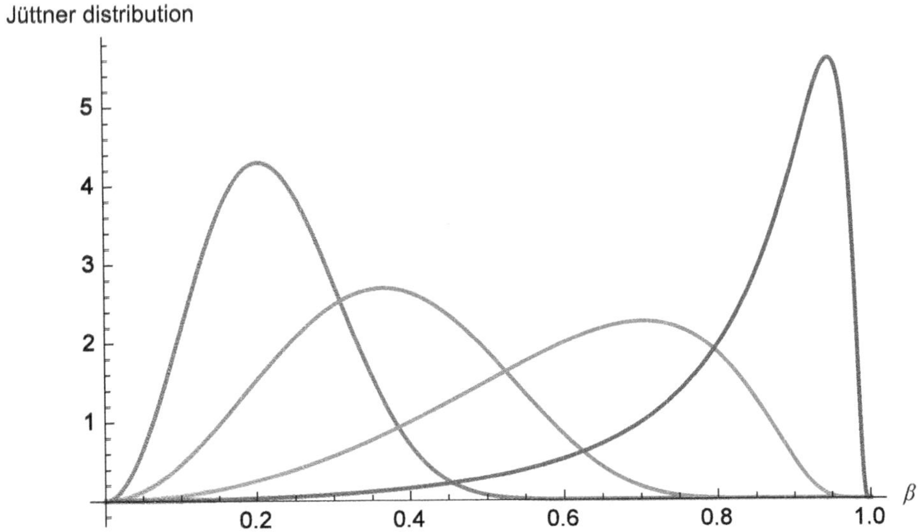


**Figure 4.1.** *Jüttner distribution [4.105], for $\theta = 0.02$ (blue), $\theta = 0.06$ (orange), $\theta = 0.2$ (green) and $\theta = 0.6$ (red). For a color version of this figure, see www.iste.co.uk/ourabah/collective.zip*

# PART 2

# Applications

# 5

# Plasmas

*A method is more important than a discovery, since the right method will lead to new and even more important discoveries.*

Lev Landau

This chapter is devoted to the application of hydrodynamic and kinetic methods, discussed in Chapters 1–4, to the study of plasmas. Plasmas are *ionized* media, composed of electrons, ions and, optionally, neutral atoms and other components such as dust grains. Plasmas are generally neutral, that is to say that if a sufficiently large volume of plasma were taken, the latter would generally have a neutral charge. In plasmas, particles interact through electric and magnetic fields they generate. This gives rise to collective phenomena that must be dealt with by statistical methods.

We will be particularly interested here in oscillations in plasmas, essentially electronic oscillations, although we will also study ionic oscillations. These oscillations are ubiquitous in plasmas and are the source of many phenomena.

The particular interest here in oscillations in plasmas is also motivated by the presence of similar oscillations in other media, such as gravitational systems and Bose–Einstein condensates. Understanding the physics of oscillations in plasmas will thus be a springboard for adapting these methods in the study of other oscillations in different media that can be formally processed with the same methods (see Chapters 6–8).

Plasmas are not naturally present in our immediate environment because the energy regimes to which we are accustomed are not high enough to ionize the media. On the scale of the Universe, however, plasmas are everywhere and account for about 99% of ordinary matter (i.e. outside of dark matter and dark energy). They are found especially in interstellar media, stars, gas nebulae, quasars and pulsars, comet tails, etc. Closer to us, plasmas are present, for example, in the solar corona (see Figure 5.1). They are also present in the trail of shooting stars, the northern lights and the heart of flames.

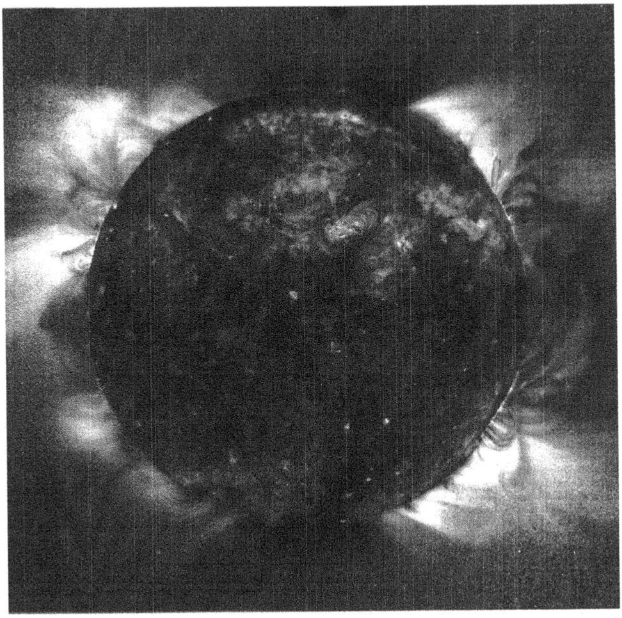

**Figure 5.1.** *The solar corona (the outermost layer of the Sun's atmosphere) represents a plasma (source: SOHO image, ESA/NASA). For a color version of this figure, see www.iste.co.uk/ourabah/collective.zip*

Plasmas can also be produced in the laboratory and are the source of many technologies. They are particularly present in electric discharges and find applications, for example, in plasma propulsion. They are also used in nuclear fusion, for example in *tokamaks*, that is, magnetic confinement devices exploring the possibility of producing energy by nuclear fusion (see Figure 5.2).

Thus, a description of the collective phenomena operating in plasmas, with statistical tools, is necessary both to understand various astrophysical or laboratory phenomena and control them for technological and industrial purposes.

**Figure 5.2.** *Diagram of the ITER Tokamak, currently the world's largest tokamak (source: ITER). For a color version of this figure, see www.iste.co.uk/ourabah/collective.zip*

## 5.1. Electronic oscillations in classical plasmas: a hydrodynamic approach

Consider a conventional non-magnetized plasma that will be described here using the hydrodynamic approach (see Chapter 2). For the sake of simplicity, we will assume immobile ions (as ions are much more massive than electrons, their movement may be neglected) and thus study only the dynamics of electrons. Also, at first, we will not consider the pressure term (which amounts to assuming a zero temperature). Such an approach is, indeed, unrealistic, but it will allow us to define a time scale that plays a crucial role in plasma physics, namely the *plasma frequency*.

With the above assumptions, the system of fluid equations to consider is as follows:

$$\begin{aligned}
&\frac{\partial n_e}{\partial t} + \nabla \cdot (n_e \mathbf{v}_e) = 0 \\
&m n_e \left( \frac{\partial \mathbf{v}_e}{\partial t} + \mathbf{v}_e \nabla \mathbf{v}_e \right) = -e n_e \mathbf{E} \\
&\varepsilon_0 \nabla \cdot \mathbf{E} = \varepsilon_0 \partial \mathbf{E} / \partial \mathbf{x} = e\,(n_i - n_e)
\end{aligned} \qquad [5.1]$$

where the first two equations represent, respectively, the continuity equation and the Euler equation, while the last equation represents the Maxwell–Gauss equation (also called, by malapropism, a Poisson equation, in the context of plasma physics).

We are thus dealing with a system of partial nonlinear differential equations, for which an exact solution is not possible. We will apply a disruptive method to have an approximate solution. Let us consider small perturbations around the equilibrium values of density $n_0$, velocity $v_0$ and electric field $\mathbf{E_0}$, namely:

$$\begin{aligned} n_e &= n_0 + n_1 \\ \mathbf{v_e} &= \mathbf{v_0} + \mathbf{v_1} \\ \mathbf{E} &= \mathbf{E_0} + \mathbf{E_1} \end{aligned} \qquad [5.2]$$

Since the plasma is almost neutral and at equilibrium before the perturbation, we have:

$$\nabla n_0 = \mathbf{v}_0 = \mathbf{E}_0 = 0\,; \quad \frac{\partial n_0}{\partial t} = \frac{\partial \mathbf{v}_0}{\partial t} = \frac{\partial \mathbf{E}_0}{\partial t} = 0 \qquad [5.3]$$

These conditions express the fact that the state of equilibrium is *stationary*, that is to say not dependent on time, and that the plasma is initially immobile, without electrical disturbance and homogeneous density.

By injecting relationships [5.2] into the system of fluid equations [5.1] and keeping only the terms of order 1 (linearization), system [5.1] then becomes:

$$\begin{aligned} & m_e \frac{\partial \mathbf{v}_1}{\partial t} = -e\mathbf{E_1} \\ & \varepsilon_0 \nabla \mathbf{E}_1 = e n_1 \\ & \frac{\partial n_1}{\partial t} + n_0 \nabla \cdot \mathbf{v_1} = 0 \end{aligned} \qquad [5.4]$$

We are looking for solutions of the *plane wave* type[1] of wave vector **k**, and frequency $\omega$, that is:

$$n_1 \sim e^{i(kx-\omega t)} \qquad [5.5]$$

1 More formally, this amounts to making a Fourier transform. For a system of equations with linear partial derivatives (as for the equations [5.4]), a Fourier transform amounts to looking for a solution in the form of a plane wave.

where we have assumed, without losing generality, that the wave vector $\mathbf{k}$ is parallel to the $x$ axis. We then have, by combining equations [5.4]:

$$\frac{\partial^2 n_1}{\partial t} - \omega_p^2 n_1 = 0 \qquad [5.6]$$

where we defined:

$$\omega_p \equiv \left(\frac{n_0 e^2}{\varepsilon_0 m_e}\right)^{1/2} \qquad [5.7]$$

Equation [5.6] indicates that, under these conditions, especially in the absence of pressure, electrons oscillate at an eigenfrequency $\omega_p$ [5.7], which is called *plasma frequency*. Due to its dependence on the density at equilibrium alone, the latter represents an appropriate unit of time (more precisely, its inverse $1/\omega_p$ plays the role of unit of time). Note that in this case, the group velocity $v_g = d\omega/dk$ is zero. It is then said that the medium is not *dispersive*.

This approach, which ignores pressure forces, is only valid within the hypothetical zero temperature limit. Thermal agitation, in fact, allows plasma oscillation to propagate like a wave. To see this, we must consider the effect of the pressure force ($-\nabla p_e$) in the Euler equation. Consider an isothermal pressure term of the *ideal gas*, that is:

$$p_e = 3k_B T_e n_e \qquad [5.8]$$

where temperature will be considered constant (and assumed to be known). Thus, we have:

$$\nabla p_e = 3k_B T_e \nabla n_e = 3k_B T_e \nabla (n_0 + n_1) = 3k_B T_e \frac{\partial n_1}{\partial x} \hat{\mathbf{x}} \qquad [5.9]$$

Then, the linearized Euler equation becomes:

$$m_e n_0 \frac{\partial \mathbf{v_1}}{\partial t} = -en_0 \mathbf{E_1} - 3k_B T_e \nabla n_1 \qquad [5.10]$$

Considering perturbations in the form of plane waves (with $\mathbf{k}$ parallel to the $x$ axis) here too, we have:

$$-im_e \omega n_0 v_1 = -en_0 E_1 - 3k_B T_e ikn_1 \qquad [5.11]$$

The continuity equation and the Maxwell–Gauss equation are given by:

$$- i\omega n_1 = -n_0 ikv_1$$
$$ik\varepsilon_0 E_1 = -en_1 \qquad [5.12]$$

This allows us to get:

$$im_e\omega n_0 v_1 = \left[en_0\left(\frac{-e}{ik\varepsilon_0}\right) + 3k_BT_e ik\right]\frac{n_0 ik}{i\omega}v_1 \qquad [5.13]$$

or even:

$$\omega^2 v_1 = \left(\frac{n_0e^2}{\varepsilon_0 m_e} + \frac{3k_BT_e}{m}k^2\right)v_1 \qquad [5.14]$$

or finally a dispersion relation of electronic oscillations (or Langmuir waves) in the form:

$$\omega^2 = \omega_p^2 + \frac{3}{2}k^2v_{th}^2 \qquad [5.15]$$

where $v_{th}^2 = 2k_BT_e/m_e$. We see that frequency $\omega$ now depends on the wavenumber $k$. The medium is said to be dispersive because the phase velocity $v_g$ is no longer zero in this case. Indeed, we have:

$$v_g = \frac{d\omega}{dk} = \frac{3}{2}\frac{k}{\omega}v_{th}^2 = \frac{3}{2}\frac{v_{th}^2}{v_\phi} \qquad [5.16]$$

Note that the dispersion relation [5.15] can be rewritten in the dimensionless form:

$$\omega^2/\omega_p^2 = 1 + 3K^2 \qquad [5.17]$$

where we have defined $K \equiv \lambda_D k$, where:

$$\lambda_D = v_{th}/\omega_p = \sqrt{\frac{\varepsilon_0 k_BT_e}{n_0e^2}} \qquad [5.18]$$

is called the *Debye length*.

Dispersion relation [5.17] is illustrated in Figure 5.3. It can be seen that frequency $\omega$ coincides with eigenfrequency $\omega_p$ only for $K = \lambda_D k = 0$, for a zero temperature (see relation [5.18]).

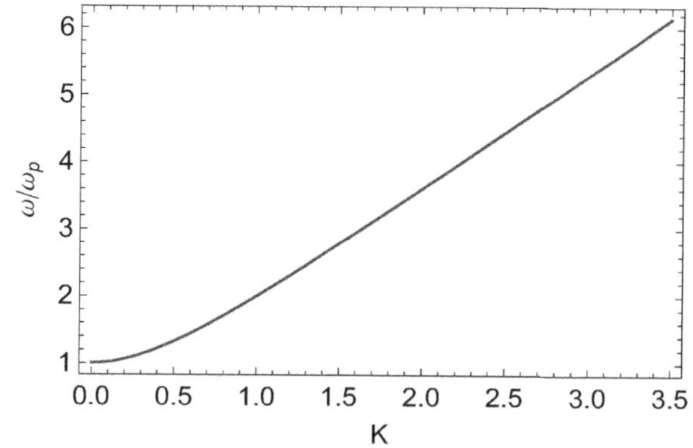


**Figure 5.3.** *Dispersion relation [5.17]*

## 5.2. Ion waves: hydrodynamic approach

So far, we have only considered the dynamics of electrons. Since ions are much more massive than electrons, their dynamics can be ignored. However, their dynamics are essential in the propagation of low-frequency waves.

To understand this phenomenon, it may be useful to recall the propagation of sound waves in a neutral gas, for example, in air. By adopting a fluid description, the hydrodynamic model can be written as follows:

$$\begin{aligned} &\frac{\partial \rho}{\partial t} + \nabla \cdot (\rho \mathbf{v}) = 0 \\ &\rho \left[ \frac{\partial v}{\partial t} + (\mathbf{v} \cdot \nabla)\mathbf{v} \right] = -\nabla p = -\frac{\gamma p}{\rho} \nabla \rho \end{aligned} \qquad [5.19]$$

where $\rho = mn_0$ is the mass density (density), the only force present is the pressure force $-\nabla p$, and $\gamma$ is the adiabatic index of the gas[2]. By linearizing around equilibrium values $\rho_0$ and $p_0$, and by searching for solutions in the form of plane waves, we have:

$$\begin{aligned} &-i\omega \rho_0 \mathbf{v}_1 = -\frac{\gamma p_0}{\rho_0} i\mathbf{k} \rho_1 \\ &-i\omega \rho_1 + \rho_0 i\mathbf{k} \cdot \mathbf{v}_1 = 0 \end{aligned} \qquad [5.20]$$

2 For an ideal monoatomic gas, $\gamma$ is always equal to $5/3$. For an ideal diatomic gas, it is $7/5 \approx 1.4$ under usual temperature conditions.

Considering **k** and $\mathbf{v_1}$ parallel to the $x$ axis, we have:

$$-i\omega\rho_0 v_1 = -\frac{\gamma p_0}{\rho_0} ik \frac{\rho_0 ikv_1}{i\omega}$$
$$\omega^2 v_1 = k^2 \frac{\gamma p_0}{\rho_0} v_1 \qquad [5.21]$$

From this, it is possible to deduce the velocity of sound $c_s$ in the neutral gas as follows:

$$\frac{\omega}{k} = \left(\frac{\gamma p_0}{\rho_0}\right)^{1/2} = \left(\frac{\gamma k_B T}{m}\right)^{1/2} = c_s \qquad [5.22]$$

In a low-collision plasma with a small population of neutral atoms, a similar phenomenon appears; this is called the *ion acoustic wave* or simply *ion sound wave.*

To study the ion acoustic wave, let us write the Euler equation of the fluid model for ions:

$$m_i n_i \left[\frac{\partial \mathbf{v_i}}{\partial t} + (\mathbf{v_i} \cdot \nabla)\,\mathbf{v}_i\right] = en_i\mathbf{E} - \nabla p = -en_i\nabla\phi - \gamma_i kT_i \nabla n \qquad [5.23]$$

$m_i$ being the mass of ions and $T_i$ their temperature. By linearizing and searching for a solution in the form of plane waves, we have:

$$-i\omega m_i n_i n_0 v_{i1} = -en_0 ik\phi_1 - \gamma_i kT_i ikn_1 \qquad [5.24]$$

Since electrons are much lighter than ions, we can ignore their mass, $m_e = 0$. The electron density $n_e$ cannot then be developed around the equilibrium value $n_0$ as follows (see equation [1.85] in Chapter 1):

$$n_e = n = n_0 \exp(e\phi_1/kT_e) = n_0 (1 + e\phi_1/kT_e + \ldots) \qquad [5.25]$$

where, in the last equality, the exponential has been linearized (development limited to order 1). Perturbation in the density of electrons (and therefore of ions) is then given by:

$$n_1 = n_0 \frac{e\phi_1}{kT_e} \qquad [5.26]$$

The linearized ion continuity equation is written as:

$$i\omega n_1 = n_0 ikv_{i1} \quad [5.27]$$

By replacing $\phi_1$ and $n_1$ as functions of $v_1$, we then have:

$$i\omega m_i n_0 v_{i1} = \left( en_0 ik\frac{kT_e}{en_0} + \gamma_i kT_i ik \right) \frac{n_0 ikv_{i1}}{i\omega} \quad [5.28]$$

The dispersion relation of the ion acoustic wave is deduced therefrom as follows:

$$\omega^2 = k^2 \left( \frac{kT_e}{m_i} + \frac{\gamma_i kT_i}{m_i} \right) \quad [5.29]$$

or even:

$$\frac{\omega}{k} = \left( \frac{kT_e + \gamma_i kT_i}{m_i} \right)^{1/2} \equiv v_s \quad [5.30]$$

$v_s$ represents the speed of sound in a plasma, which is composed of electrons and ions. It is interesting to note that in the case of a zero ionic temperature ($T_i = 0$), these waves nevertheless propagate in the plasma.

## 5.3. Classic plasmas: kinetic approach

Let us now return to the problem of electronic oscillations discussed in section 5.1, but using kinetic theory. This will reveal a new phenomenon, of a purely kinetic nature, which cannot be described within the fluid model context. As in section 5.1, we consider a uniform plasma (with a velocity-dependent distribution $f_0(v)$) with no external fields $\mathbf{E_0} = \mathbf{B_0} = 0$, and we are interested in studying small perturbations around equilibrium values. Thus, the distribution function $f(\mathbf{r}, \mathbf{v}, t)$ can be broken down into an equilibrium distribution $f_0$ and a disturbed part $f_1$:

$$f(\mathbf{r}, \mathbf{v}, t) = f_0(\mathbf{v}) + f_1(\mathbf{r}, \mathbf{v}, t) \quad [5.31]$$

where the perturbed component of the distribution will be assumed to be minor. Let us write the Vlasov equation [1.25] in this case:

$$\frac{\partial f}{\partial t} + \mathbf{v} \cdot \nabla f - \frac{e\mathbf{E}}{m} \cdot \frac{\partial f}{\partial \mathbf{v}} = 0 \quad [5.32]$$

Considering [5.31], it can therefore be linearized (keeping only the terms of order 1) to be written in the form:

$$\frac{\partial f_1}{\partial t} + \mathbf{v} \cdot \nabla f_1 - \frac{e}{m}\mathbf{E}_1 \cdot \frac{\partial f_0}{\partial \mathbf{v}} = 0 \qquad [5.33]$$

Considering perturbations in the form of plane waves (with a wave vector **k** assumed to be parallel to the x axis) here too, we have:

$$f_1 \sim e^{i(kx-\omega t)} \qquad [5.34]$$

and the linearized Vlasov equation [5.33] gives:

$$\begin{aligned} -i\omega f_1 + ikv_x f_1 &= \frac{e}{m}E_x \frac{\partial f_0}{\partial v_x} \\ f_1 &= \frac{ieE_x}{m}\frac{\partial f_0/\partial v_x}{\omega - kv_x} \end{aligned} \qquad [5.35]$$

Using the fact that the integral of the distribution over the space of velocities gives a density (see equation [1.19] in Chapter 1), the Maxwell–Gauss equation gives:

$$\begin{aligned} \varepsilon_0 \nabla \cdot \mathbf{E}_1 &= ik\varepsilon_0 E_x = -en_1 \\ &= -e\iiint f_1 d^3v \end{aligned} \qquad [5.36]$$

Using the expression for $f_1$, we have:

$$1 = -\frac{e^2}{km\varepsilon_0}\iiint \frac{\partial f_0/\partial v_x}{\omega - kv_x} d^3v \qquad [5.37]$$

If we replace $f_0$ by a function $\hat{f}_0$ normalized to 1 (i.e. $\hat{f}_0 = f_0/n_0$), we have:

$$1 = -\frac{\omega_p^2}{k}\int_{-\infty}^{\infty} dv_z \int_{-\infty}^{\infty} dv_z \int_{-\infty}^{\infty} \frac{\partial \hat{f}_0(v_x, v_y, v_z)/\partial v_x}{\omega - kv_x} dv_x \qquad [5.38]$$

If the plasma is initially at equilibrium, $f_0$ is a three-dimensional Maxwell–Boltzmann type function. After implementing the first two integrals, all that remains is a one-dimensional function $f_0(v_x)$ and the dispersion relation then becomes:

$$1 = \frac{\omega_p^2}{k^2}\int_{-\infty}^{\infty} \frac{\partial \hat{f}_0(v_x)/\partial v_x}{v_x - (\omega/kv)} dv_x \qquad [5.39]$$

Integral [5.39] is not well defined because of the singularity. It should be treated as a contour integral in the complex velocity plane. Russian physicist Landau (1946) was the first to treat this integral, thus revealing a new phenomenon now called *Landau damping*. By using integral transformation techniques, Landau indeed showed that dispersion relation [5.39] can be reduced to:

$$1 = \frac{\omega_p^2}{k^2}\left[P\int_{\infty}^{\infty}\frac{\partial \hat{f}_0/\partial v}{v-(\omega/k)}dv + i\pi\frac{\partial \hat{f}_0}{\partial v}\bigg|_{v=w/k}\right] \quad [5.40]$$

where we set $v \equiv v_x$. $P$ represents the *principal Cauchy value*. It corresponds to an integration along the real axis excluding the region around the pole.

The second term in equation [5.40], on the other hand, is due to the pole and produces an imaginary contribution to the frequency.

First, let us deal with the real part of the dispersion relation [5.40], looking only at the principal Cauchy value for the moment.

The integral can be evaluated by following an integration per part as follows:

$$\begin{aligned}\int_{-\infty}^{\infty}\frac{\partial \hat{f}_0}{\partial v}\frac{dv}{v-(\omega/k)} &= \left[\frac{\hat{f}_0}{v-v_\phi}\right]_{-\infty}^{\infty} - \int_{-\infty}^{\infty}\frac{-\hat{f}_0 dv}{\left(v-v_\phi\right)^2} \\ &= \int_{-\infty}^{\infty}\frac{\hat{f}_0 dv}{\left(v-v_\phi\right)^2}\end{aligned} \quad [5.41]$$

where $v_\phi = \omega/k$ is the phase velocity. We thus simply have the mean value of $\left(v-v_\phi\right)^{-2}$ according to distribution $\hat{f}_0$. The real part of the dispersion relation is then given by:

$$1 = \frac{\omega_p^2}{k^2}\langle\left(v-v_\phi\right)^{-2}\rangle \quad [5.42]$$

Consider the case of a high phase velocity $v << v_\phi$. In this case, we can do a Taylor series development to have:

$$\left(v-v_\phi\right)^{-2} = v_\phi^{-2}\left(1-\frac{v}{v_\phi}\right)^{-2} \approx v_\phi^{-2}\left(1+\frac{2v}{v_\phi}+\frac{3v^2}{v_\phi^2}+\frac{4v^3}{v_\phi^3}\right) \quad [5.43]$$

It may be noted that, since the distribution $\hat{f}_0$ is even, the odd terms disappear when the mean is taken. Then, we simply have:

$$\langle (v - v_\phi)^{-2} \rangle = v_\phi^{-2} \left( 1 + \frac{3\langle v^2 \rangle}{v_\phi^2} \right) \qquad [5.44]$$

For a Maxwell–Boltzmann distribution, we have:

$$\frac{1}{2} m \langle v_x^2 \rangle = \frac{1}{2} k_B T_e \qquad [5.45]$$

The real part of the dispersion relation then becomes:

$$1 = \frac{\omega_p^2}{k^2} \frac{k^2}{\omega^2} \left( 1 + 3 \frac{k^2}{\omega^2} \frac{k_B T_e}{m} \right) \qquad [5.46]$$

or even:

$$\omega^2 = \omega_p^2 + \frac{\omega_p^2}{\omega^2} \frac{3 k_B T_e}{m} k^2 \qquad [5.47]$$

We thus have the result obtained in section 5.1 using the fluid approach [5.15], namely:

$$\omega^2 = \omega_p^2 + \frac{3 k_B T_e}{m} k^2 \qquad [5.48]$$

Note however that the use of the kinetic theory has made it possible to show the existence of a pole in the integral relation [5.39], thus giving an imaginary contribution to the dispersion relation. To evaluate the effect of the imaginary part in the dispersion relation, the thermal correction on the real part is initially neglected. Using a limited expansion, dispersion relation [5.40] then becomes:

$$1 = \frac{\omega_p^2}{\omega^2} + i\pi \frac{\omega_p^2}{k^2} \frac{\partial \hat{f}_0}{\partial v} \bigg|_{v=v_\phi} \qquad [5.49]$$

or even:

$$\omega_p^2 = \omega^2 \left( 1 - i\pi \frac{\omega_p^2}{k^2} \left[ \frac{\partial \hat{f}_0}{\partial v} \right]_{v=v_\phi} \right) \qquad [5.50]$$

Assuming that the imaginary part is small in comparison with the real part, we have:

$$\omega^2 = \omega_p^2 \left( 1 + i\pi \frac{\omega_p^2}{k^2} \left[ \frac{\partial \hat{f}_0}{\partial v} \right]_{v=v_\phi} \right) \quad [5.51]$$

In particular, if the equilibrium distribution $\hat{f}_0$ is given by the Maxwell–Boltzmann distribution, we have:

$$\begin{aligned} \frac{\partial \hat{f}_0}{\partial v} &= \left(\pi v_{th}^2\right)^{-1/2} \left( \frac{-2v}{v_{th}^2} \right) \exp\left( \frac{-v^2}{v_{th}^2} \right) \\ &= -\frac{2v}{\sqrt{\pi} v_{th}^3} \exp\left( \frac{-v^2}{v_{th}^2} \right) \end{aligned} \quad [5.52]$$

Taking $v_\phi = \omega_p / k$ in the coefficient but keeping the thermal correction in the exponent, we have:

$$Im(\omega) = -\frac{\pi}{2} \frac{\omega_p^3}{k^2} \frac{2\omega_p}{k\sqrt{\pi}} \frac{1}{v_{th}^3} \exp\left( \frac{-\omega^2}{k^2 v_{th}^2} \right) \quad [5.53]$$

Thus, there is an imaginary contribution to the dispersion relation. To interpret the physical effect of the imaginary part, we must remember that to arrive at this point, we had considered perturbations in the form of plane waves, the temporal part of which gives $\sim e^{-i\omega t}$. An imaginary contribution to the frequency thus produces an exponential evolution with time, that is $\sim e^{\gamma t}$, where $\gamma$ is the imaginary part of the frequency. Since the latter is negative in this case (see [5.53]), it produces a decreasing exponential over time, that is a damping.

Beyond the mathematical tools presented here, it is above all important to understand the physical origin of this result. Recall that the presence of damping is generally related to the presence of collisions. Thus, if a pendulum that oscillates in the air, for example, loses its energy, it is precisely because of collisions with molecules that, in the long term, take its energy: this is called *dissipation*. Our starting point here was the Vlasov equation that assumes there are no collisions. What is the origin of this damping?

Its origin is actually the exchange of energy that occurs between particles and the wave. Indeed, particles with a velocity close to the phase velocity of the wave exchange energy optimally with the wave (these particles are said to be resonant). Particles with a velocity greater than $v_\phi$ yield part of their energy to the wave and particles with a velocity less than $v_\phi$ take energy from the wave. With the distribution being normalized, there are always more particles with velocities $v < v_\phi$ (thus taking energy from the wave) than particles with a velocity $v > v_\phi$ (thus giving energy to the wave). It is this asymmetry that produces, over the long term, a damping of the wave (see Figure 5.4). Landau damping is of significant importance in plasmas. For example, it is used to generate currents in tokamaks. Note, however, that it is not unique to plasma physics; it also appears in galactic dynamics where the gas of electrons interacting through electrical forces is replaced by a "stellar gas" interacting through gravitational forces.

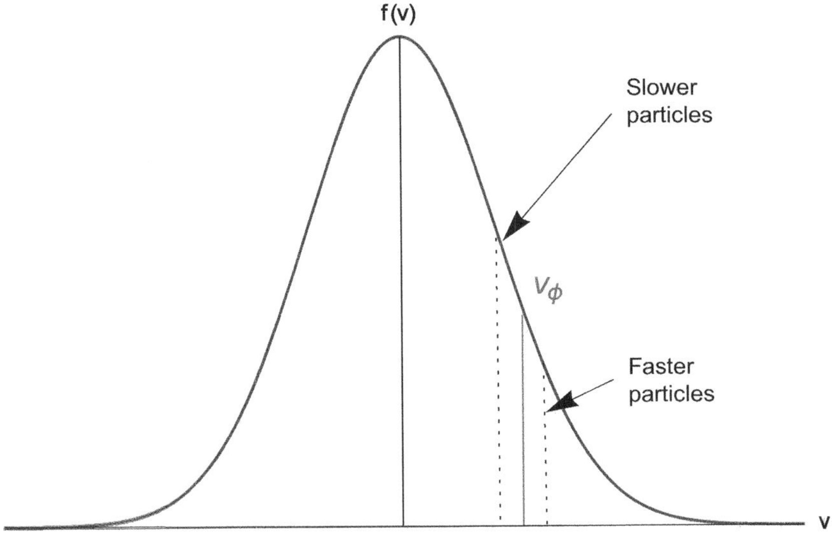


**Figure 5.4.** *Illustration of the kinetic origin of Landau damping. Particles with velocities lower than the phase velocity $v_\phi$ (thus taking a part of the energy of the wave) are more numerous than particles with a velocity higher than the phase velocity. For a color version of this figure, see www.iste.co.uk/ourabah/collective.zip*

Finally, it should be noted that there is a phenomenon similar to Landau damping, but where the exchange of energy takes place in the opposite direction: the particles of the medium give energy to the wave. This is the *Cherenkov radiation* that occurs when a charged particle moves through a dielectric medium at a speed greater than the speed of light in that medium. It is this effect that produces the bluish luminosity of water surrounding the core of nuclear reactors.

## 5.4. Quantification of electronic oscillations: hydrodynamic and kinetic approaches

Let us return to the fluid approach to electronic oscillations, discussed in section 5.1, but this time using the quantum hydrodynamic model developed in Chapter 3. As mentioned in section 5.1, we will initially ignore the effect of pressure. The only difference, in comparison with the classic hydrodynamic model, is the presence of the Bohm potential $Q$ in the Euler equation (see equation [3.10] in Chapter 3). So, we get:

$$\begin{aligned} &\frac{\partial n_e}{\partial t} + \nabla \cdot (n_e \mathbf{v}_e) = 0 \\ &m n_e \left( \frac{\partial \mathbf{v}_e}{\partial t} + \mathbf{v}_e \nabla \mathbf{v}_e \right) = -e n_e \mathbf{E} + \frac{\hbar^2}{2m_e} \nabla \left( \frac{\nabla^2 \sqrt{n_e}}{\sqrt{n_e}} \right) \\ &\varepsilon_0 \nabla \cdot \mathbf{E} = \varepsilon_0 \partial \mathbf{E} / \partial \mathbf{x} = e\,(n_i - n_e) \end{aligned} \quad [5.54]$$

where the first equation represents the continuity equation and the second represents the Euler equation in the presence of the Bohm potential, while the last equation represents the Maxwell–Gauss equation. Following the same steps as in section 5.1, considering small perturbations around the equilibrium values, $n_e = n_0 + n_1$, $\mathbf{v_e} = \mathbf{v_0} + \mathbf{v_1}$ and $\mathbf{E} = \mathbf{E_0} + \mathbf{E_1}$, and looking for solutions in the form of plane waves, we get:

$$\omega^2 = \omega_p^2 + \frac{\hbar^2 k^4}{4m_e^2} \quad [5.55]$$

Thus, even in the absence of pressure forces, a form of dispersion, of quantum origin, exists in the medium (the eigenfrequency $\omega_p$ being found only within the classic limit, $\hbar = 0$). In this case, the group velocity is given by:

$$v_g \equiv \frac{d\omega}{dk} = \frac{\hbar^2 k^3}{m_e^2} \quad [5.56]$$

The effect of pressure forces can also be included in the hydrodynamic model [5.54]. However, since these forces are independent of quantum effects, what would be obtained would simply be the classic dispersion relation [5.15] to which the quantum correction $\hbar^2 k^4 / 4m_e^2$ will be added, namely:

$$\omega^2 = \omega_p^2 + \frac{3k_B T_e}{m} k^2 + \frac{\hbar^2 k^4}{4m_e^2} \quad [5.57]$$

It should be noted, however, that a more realistic model would take into account the fermionic nature of electrons and the effect of Fermi–Dirac statistics. The latter is relatively difficult to manipulate because these moments cannot be calculated in a simple manner as in the case of the Maxwell–Boltzmann distribution. Thus, the exact form of the pressure term in the quantum case is quite difficult to evaluate and will not be discussed here.

Let us now examine the same problem in the kinetic framework, through the Wigner–Moyal kinetic theory presented in Chapter 3. In this case, the Wigner–Moyal kinetic equation [3.58] is written as:

$$\left(\frac{\partial}{\partial t}+\frac{\hbar}{m}\mathbf{k}\cdot\frac{\partial}{\partial \mathbf{r}}\right)W=-\frac{2e}{\hbar}W(\sin\Lambda)\phi \qquad [5.58]$$

with:

$$\Lambda=\frac{1}{2}\frac{\overleftarrow{\partial}}{\partial \mathbf{k}}\cdot\frac{\overrightarrow{\partial}}{\partial \mathbf{r}} \qquad [5.59]$$

where we have associated potential $V$ with the electric potential, that is, $V\equiv -e\phi$. Equation [5.58] will have to be coupled with the Poisson equation, thus linking potential $\phi$ to density, and therefore to the distribution function $W$. To use the same notations as in section 5.3, note $W\to f$. Expanding on equation [5.58], we have (in one dimension):

$$\frac{\partial f}{\partial t}+v\frac{\partial f}{\partial x}-\frac{e}{m}\frac{\partial \phi}{\partial x}\frac{\partial f}{\partial v}=\frac{e\hbar^2}{24m^3}\frac{\partial^3\phi}{\partial x^3}\frac{\partial^3 f}{\partial v^3}+O\left(\hbar^4\right) \qquad [5.60]$$

Thus, it is clear that the Vlasov equation is found within the classic limit $\hbar\to 0$. The Wigner–Moyal equation must be coupled with the Poisson equation:

$$\frac{\partial^2\phi}{\partial x^2}=\frac{e}{\varepsilon_0}\left(\int f dv-n_0\right) \qquad [5.61]$$

where we assumed ions to be immobile, with a density $n_i\equiv n_0$. Following the same steps as in section 5.3, the following dispersion relation is obtained:

$$1+\frac{m\omega_p^2}{n_0 k}\int\frac{f_0(v+\hbar k/2m)-f_0(v-\hbar k/2m)}{\hbar k(\omega-kv)}dv=0 \qquad [5.62]$$

By separating the integral and changing variables, the integral can be rewritten as follows:

$$1-\frac{\omega_p^2}{n_0}\int\frac{f_0(v)}{(\omega-kv)^2-\hbar^2k^4/4m^2}dv=0 \qquad [5.63]$$

In this form, it is clear that within the classic limit $\hbar\to 0$, we have:

$$1+\frac{\omega_p^2}{n_0k}\int\frac{\partial f_0/\partial v}{\omega-kv}dv=0 \qquad [5.64]$$

which is none other than the classic dispersion relation [5.39].

The presence of the pole makes the [5.63] type of integrals quite difficult to handle. As in the classic case, however, the Cauchy principal value of integral [5.63] gives the dispersion relation obtained in the context of the quantum hydrodynamic model [5.55]. The correction due to the pole produces a Landau damping. Processing it in the quantum context is quite subtle and will not be further discussed here. The physical origins of this phenomenon, discussed in the classic context, nevertheless remain valid in the quantum context.

## 5.5. In a relativistic plasma: kinetic approach

Let us now return to the case of a collisionless plasma processed kinetically, but this time within the relativity context. Thus, our system consists of a set of $N$ charged relativistic particles interacting via electromagnetic interactions. It may be useful in this case to transform the stress-energy four-vector such as:

$$p^\mu\longrightarrow p^\mu-\frac{q}{c}A^\mu \qquad [5.65]$$

where $A^\mu$ is the vector potential of the electromagnetic field. The equations of momentum are given by:

$$\frac{dp^\mu}{d\tau}=\frac{q}{c}F^{\mu\nu}u_\nu,\quad p^\mu=mu^\mu,\quad u^\mu=\frac{dx^\mu}{d\tau} \qquad [5.66]$$

where $F^{\mu\nu}$ is the electromagnetic field tensor and $u^\mu$ is the four-velocity of the particle. This momentum equation can be written using a Fourier transform of the field as follows:

$$\frac{dp^\mu}{d\tau}=\frac{q}{c}F^{\mu\nu}u_\nu=\frac{iq}{m}\int\frac{d^4k'}{(2\pi)^4}\exp[-ikx(\tau)]\left(k'u\right)G^{\mu\nu}\left(k',u\right)A_\nu\left(k'\right) \qquad [5.67]$$

where:

$$G^{\mu\nu}(k,u) = g^{\mu\nu} - \frac{k^\mu u^\nu}{ku} \tag{5.68}$$

As indicated above, a Fourier transform amounts to assuming perturbations in the form of plane waves, that is:

$$A^\mu(x) \propto \exp\left(-ik^\mu x_\mu\right) \equiv \exp(-ikx) = \exp[-i(\omega t - \mathbf{kx})] \tag{5.69}$$

where $k$ represents the four-vector $k \equiv k^\mu = (\omega/c, \mathbf{k})$, $\omega$ is the frequency and $\mathbf{k}$ is the wave vector. Considering a perturbation of the distribution, we can write:

$$\mathscr{F}(k,p) = \mathscr{F}(p)(2\pi)^4\delta^4(k) + \mathscr{F}_1(k,p) \tag{5.70}$$

where $\mathscr{F}_1(k,p)$ is a supposedly small perturbation. In the absence of collisions, distribution [5.70] satisfies the relativistic Vlasov equation (equation [4.52] without collision term) which, in this case, is written as:

$$\left[-ip^\mu k_\mu + m\frac{dp^\mu}{d\tau}\frac{\partial}{\partial p^\mu}\right]\mathscr{F}(k,p) = 0 \tag{5.71}$$

where $dp^\mu/d\tau$ is given by relationship [5.66]. The solution for the disturbed distribution is then given by:

$$\mathscr{F}_1(k,p) = qG^{\mu\nu}(k,u)A_\nu(k)\frac{\partial \mathscr{F}(p)}{\partial p^\mu} \tag{5.72}$$

Recall that the four-current is given by:

$$J_1^\mu(k) = c\sum q\int \mathscr{F}_1(k,p)p^\mu d^4p \tag{5.73}$$

It may be useful to separate the current into two components: an *induced* component and another *external* component:

$$J^\mu(k) = J^\mu_{ind}(k) + J^\mu_{ext}(k) \tag{5.74}$$

If the induced current is sufficiently low, we have:

$$4\pi J^\mu_{ind}(k) = \Pi^\mu_\nu(k)A^\nu(k) \tag{5.75}$$

where $\Pi^{\mu}_{\nu}$ is called the *linear response tensor*. Using the four-vector current [5.73], we can generally write:

$$\Pi^{\mu\nu}(k) = c\sum \frac{q^2}{m} \int d^4p G^{\lambda\nu}(k,u) p^{\mu} \frac{\partial \mathscr{F}(p)}{\partial p^{\lambda}} \quad [5.76]$$

Here, we will limit ourselves to the case of an *isotropic* plasma in its Lorentz reference frame. In this reference frame, we get fluid velocity from $U^{\mu} = (c,0,0,0)$. In this case, the derivative of the distribution with respect to the energy-stress four-vector is reduced to a derivative with respect to the energy of the particles in the Lorentz reference frame. Thus, we have:

$$\Pi^{\mu\nu}(k) = c\sum \frac{q^2}{m} \int d^4p \left[(uU)U^{\mu}U^{\nu} - \frac{(kU)u^{\mu}u^{\nu}}{(ku)}\right] \frac{\partial \mathscr{F}(p)}{\partial (pU)} \quad [5.77]$$

In the isotropic case, the linear response tensor can be broken down into three components: a *longitudinal* part, a *transverse* part and a *rotational* part. The latter cancels out in the case of a conventional plasma (Vereshchagin and Aksenov 2017). Therefore, we have:

$$\Pi^{\mu\nu}(k) = \Pi^{L}(k)L^{\mu\nu}(k,U) + \Pi^{T}(k)T^{\mu\nu}(k,U) \quad [5.78]$$

where tensors $L^{\mu\nu}(k,U)$ and $T^{\mu\nu}(k,U)$ are built from variables of four-vectors $k^{\mu}$ and $U^{\mu}$ as follows:

$$\begin{aligned} L^{\mu\nu} &= -L^{\mu}L^{\nu} \\ T^{\mu\nu} &= \frac{(kU)^2}{k^2}L^{\mu}L^{\nu} + g^{\mu\nu} - \frac{k^{\mu}k^{\nu}}{k^2} \\ L^{\mu} &= \frac{k_{\lambda}G^{\lambda\mu}}{[k^2 - (kU)^2]^{1/2}} \end{aligned} \quad [5.79]$$

Note that from Maxwell's equations, we can write:

$$\Lambda^{\mu\nu}(k)A_{\nu}(k) \equiv \left[k^2 g^{\mu\nu} - k^{\mu}k^{\nu} + \Pi^{\mu\nu}(k)\right] A_{\nu}(k) = -4\pi J^{\mu}_{ext}(k) \quad [5.80]$$

where, in the same way, tensor $\Lambda^{\mu\nu}(k)$ can be broken down as follows:

$$\Lambda^{\mu\nu}(k) = \Lambda^{L}(k)L^{\mu\nu}(k,U) + \Lambda^{T}(k)T^{\mu\nu}(k,U) \quad [5.81]$$

where:

$$\Lambda^L(k) = \frac{(kU)^4}{k^4} L_{\mu\nu}\Lambda^{\mu\nu}, \quad \Lambda^T(k) = \frac{1}{2} T_{\mu\nu}\Lambda^{\mu\nu} \qquad [5.82]$$

The dispersion relation is obtained from the condition:

$$\Lambda^L(k) = \left[\Lambda^T(k)\right]^2 = 0 \qquad [5.83]$$

We then have the following general form of the dispersion relation:

$$(kU)^2 + \Pi^L(k) = 0 \quad k^2 + \Pi^T(k) = 0 \qquad [5.84]$$

describing longitudinal (L) and transverse (T) waves.

We can now analyze relationship [5.76] more specifically in the case of a thermal plasma. Consider a thermal plasma where the distribution at equilibrium is given by the Jüttner distribution (see [4.86]):

$$f(p) = \frac{n}{4\pi(mc)^3\theta K_2\left(\theta^{-1}\right)} \exp\left[\frac{\phi - p^\mu U_\mu}{k_B T}\right] \qquad [5.85]$$

Using equilibrium distribution [5.85], we can do an angular integration of equation [5.76]. Then, we get (Vereshchagin and Aksenov 2017):

$$\begin{aligned}
&\Pi^L(k) = -\omega_p^2\Phi^L(z,\theta), \quad \Pi^T(k) = -\omega_p^2\Phi^T(z,\theta) \\
&\Phi^L(z,\theta) = -\frac{z^2}{\theta} + \frac{z^3}{2\theta^2 K_2\left(\theta^{-1}\right)} \int_0^\infty d\chi \sinh\chi \cosh^2\chi \mathscr{A}(\theta,\chi) \qquad [5.86] \\
&\Phi^T(z,\theta) = -\frac{z^2}{2\theta} + \frac{z}{4\theta^2 K_2\left(\theta^{-1}\right)} \int_0^\infty d\chi \sinh\chi \left(z^2\cosh^2\chi - \sinh^2\chi\right) \mathscr{A}(\theta,\chi)
\end{aligned}$$

with:

$$\mathscr{A}(\theta,\chi) = \exp\left(-\frac{\cosh\chi}{\theta}\right) \ln\left(\frac{z\cosh\chi + \sinh\chi}{z\cosh\chi - \sinh\chi}\right) \qquad [5.87]$$

where variables $z$ and $\chi$ are given by:

$$\begin{aligned}
z &= -\frac{(kU)}{\left[k^2 - (kU)^2\right]^{1/2}} \\
\cosh\chi &= \frac{p^0}{\left(p^\mu p_\mu\right)^{1/2}}
\end{aligned} \qquad [5.88]$$

Quantity $z$ represents the phase velocity. In the Lorentz reference frame, it is simply reduced to $\omega/k$.

Combining equation [5.76] with the general form of the dispersion relation [5.84], we can write:

$$\omega^2 = \omega_p^2 \Phi^L(z,\theta), \quad \omega^2 = \frac{z^2 \omega_p^2}{1-z^2} \Phi^T(z,\theta) \qquad [5.89]$$

It may be useful to discuss the borderline cases of the previous relationship. This makes it possible to define the *Debye length* ($z \to 0$) as well as the *cutoff frequency* ($z \to \infty$). The Debye length can be found by analyzing the longitudinal waves, and is defined as follows:

$$\lambda_D^{-2} = \lim_{z \to 0} \frac{\Phi^L(z,\theta)}{z^2} = \frac{\omega_p^2}{c^2 \theta} \qquad [5.90]$$

This remains valid from the non-relativistic limit to the ultrarelativistic limit. As for the cutoff frequency, it should be remembered that waves with a frequency lower than the cutoff frequency cannot propagate in plasma. The cutoff frequency can be obtained, from the dispersion relation [5.87], within the limit $z \to \infty$. Within this limit, we have:

$$\omega_c^2 = \frac{\omega_p^2}{3\theta} \left\langle \left(\frac{v}{c}\right)^2 \right\rangle \to \begin{cases} \omega_p^2 \left(1 - \frac{5}{2}\theta\right), & \theta \ll 1 \\ \omega_p^2 \left(\frac{1}{3\theta} - \frac{1}{6\theta^3}\right), & \theta \gg 1 \end{cases} \qquad [5.91]$$

It may be interesting to analyze in particular Langmuir longitudinal waves or electronic oscillations ($k\lambda_D \ll 1$) in the relativistic context. In the non-relativistic limit ($\theta \ll 1$), we find the non-relativistic relation (equation [5.17]), namely:

$$\frac{\omega^2}{\omega_p^2} = 1 + 3\theta \frac{k^2 c^2}{\omega_p^2} = 1 + 3(k\lambda_D)^2 \qquad [5.92]$$

where we took the definition of Debye length [5.90] into consideration. We can also analyze ultrarelativistic behavior. For $\theta \gg 1$ and $k\lambda_D \ll 1$, on the other hand, we have:

$$\frac{\omega^2}{\omega_p^2} = \frac{1}{3\theta} \left[1 + \frac{9}{5}(k\lambda_D)^2\right] \qquad [5.93]$$

## 5.6. Beyond the linear approach: the Korteweg–De Vries equation

Until now, the waves propagating in the plasmas have been processed linearly, both in the context of the hydrodynamic model and the kinetic approach. This approximation is only valid in the case of small perturbations around equilibrium values. For larger perturbations, the approach can be extended by adopting a so-called disruptive processing. Here, we will present such a processing for the ion acoustic wave. The resulting equation, namely the Korteweg-De Vries (KdV) equation, plays a very important role in plasmas and in other media such as neutral gases and Bose–Einstein condensates.

To simplify calculations, we consider the 1D case and we will ignore the effects of pressure. The hydrodynamic model for ions is thus composed of the continuity equation:

$$\frac{\partial n_i}{\partial t}+\frac{\partial (n_i v_i)}{\partial x}=0 \tag{5.94}$$

of the Euler equation:

$$m_i\left(\frac{\partial v_i}{\partial t}+v_i\frac{\partial v_i}{\partial x}\right)=-e\frac{\partial \phi}{\partial x} \tag{5.95}$$

and Poisson equation:

$$\frac{\partial^2 \phi}{\partial x^2}=-4\pi e\left(n_i-n_e\right) \tag{5.96}$$

In order to be able to perform a perturbative processing, it is necessary to rewrite these equations in a dimensionless form. To do this, we can use the characteristic magnitudes of the plasmas, namely the Debye length $\lambda_D$, the plasma frequency $\omega_p$ and the product of the two, which defines a characteristic velocity $v_{th}=\lambda_D\omega_p$ (see equation [5.18]). Fluid equations can thus be rewritten in a dimensionless form as follows:

$$\begin{aligned}
&\frac{\partial N_i}{\partial T}+\frac{\partial (N_i V_i)}{\partial X}=0\\
&\frac{\partial V_i}{\partial T}+V_i\frac{\partial V_i}{\partial X}=-\frac{\partial \Phi}{\partial X}\\
&\frac{\partial^2 \Phi}{\partial X^2}=N_e-N_i
\end{aligned} \tag{5.97}$$

where the ion density $N_i$, fluid velocity $V_i$, electrostatic potential $\Phi$, time $T$ and space variable $X$ are normalized to $n_0$, $v_{th}$, $k_B T_e/e$ and $1/\omega_p$, respectively. Electron density, on the other hand (equation [5.25]), is written in dimensionless form as follows:

$$\begin{aligned} N_e &= \frac{n_e}{n_{e0}} = \exp(\Phi) \\ &\simeq 1 + \Phi + \frac{\Phi^2}{2} + \dots \end{aligned} \qquad [5.98]$$

where in the last stage, a development limited to the first order has been made.

Let us introduce the following variables:

$$\begin{aligned} \xi &= \varepsilon^{1/2}(X - V_p T) \\ \tau &= \varepsilon^{3/2} T \end{aligned} \qquad [5.99]$$

where $\varepsilon$ is a dimensionless parameter that measures the weakness of the amplitude and $V_p$ represents the linear phase velocity.

Derivatives of the first and second order with respect to variables of space and time are then written as:

$$\begin{aligned} &\frac{\partial}{\partial X} \to \varepsilon^{1/2}\frac{\partial}{\partial \xi}, \quad \frac{\partial^2}{\partial X^2} \to \varepsilon\frac{\partial^2}{\partial \xi^2} \\ &\frac{\partial}{\partial T} \to -\varepsilon^{1/2} V_p \frac{\partial}{\partial \xi} + \varepsilon^{3/2}\frac{\partial}{\partial \tau} \end{aligned} \qquad [5.100]$$

Equations [5.97] are then rewritten in terms of the new variables as follows:

$$\begin{aligned} &\varepsilon\frac{\partial N_i}{\partial \tau} - V_p\frac{\partial N_i}{\partial \xi} + \frac{\partial (N_i V_i)}{\partial \xi} = 0 \\ &\varepsilon\frac{\partial V_i}{\partial \tau} - V_p\frac{\partial V_i}{\partial \xi} + V_i\frac{\partial V_i}{\partial \xi} = -\frac{\partial \Phi}{\partial \xi} \\ &\varepsilon\frac{\partial^2 \Phi}{\partial \xi^2} = N_e - N_i \end{aligned} \qquad [5.101]$$

Quantities $N_i$, $V_i$, and $\Phi$ can be developed into series of powers of parameter $\varepsilon$, around equilibrium values as follows:

$$\begin{aligned}
&\Phi = 0 + \varepsilon\Phi_1 + \varepsilon^2\Phi_2 + \ldots \\
&V_i = V_0 + \varepsilon V_i^{(1)} + \varepsilon^2 V_i^{(2)} + \ldots \\
&N_i = 1 + \varepsilon N_i^{(1)} + \varepsilon^2 N_i^{(2)} + \ldots
\end{aligned} \qquad [5.102]$$

We can now substitute the latter in equations [5.101] and establish equalities for each power of $\varepsilon$.

In the smallest order in $\varepsilon$ ($\varepsilon^{3/2}$ for the first two equations and $\varepsilon$ for the Poisson equation), we have:

$$-\lambda\frac{\partial N_i^{(1)}}{\partial\xi} + \frac{\partial V_i^{(1)}}{\partial\xi} = 0 \qquad [5.103]$$

$$-\lambda\frac{\partial V_i^{(1)}}{\partial\xi} = -\frac{\partial\Phi^{(1)}}{\partial\xi} \qquad [5.104]$$

and:

$$\Phi^{(1)} - N_i^{(1)} = 0 \qquad [5.105]$$

where $\lambda = V_p - V_0$.

Then, we can deduce:

$$\begin{aligned}
&V_i^{(1)} = \frac{1}{\lambda}\Phi^{(1)} \\
&N_i^{(1)} = \frac{1}{\lambda^2}\Phi_1
\end{aligned} \qquad [5.106]$$

From the compatibility condition of the last equation with the Poisson equation, we have $\lambda = 1$, which implies:

$$V_p = 1 + V_0 \qquad [5.107]$$

By examining the higher order in $\varepsilon$ ($\varepsilon^{5/2}$ for the continuity and Euler equations and $\varepsilon^2$ for the Poisson equation), we have:

$$\begin{aligned}
&\frac{\partial N_i^{(1)}}{\partial \tau} - \lambda \frac{\partial N_i^{(2)}}{\partial \xi} + \frac{\partial V_i^{(2)}}{\partial \xi} + \frac{\partial N_i^{(1)} V_i^{(1)}}{\partial \xi} = 0 \\
&\frac{\partial V_i^{(1)}}{\partial \tau} - \lambda \frac{\partial V_i^{(2)}}{\partial \xi} + V_i^{(1)} \frac{\partial V_i^{(1)}}{\partial \xi} = -\frac{\partial \Phi^{(2)}}{\partial \xi} \\
&\frac{\partial^2 \Phi^{(1)}}{\partial \xi^2} = \frac{1}{2}\Phi^{(1)2} + \Phi^{(2)} - N^{(2)}
\end{aligned} \qquad [5.108]$$

Using these latter equations, as well as the first-order equations in $\varepsilon$ [5.106], we obtain, by rearranging the terms, the *KdV equation*:

$$\frac{\partial \Phi^{(1)}}{\partial \tau} + A\Phi^{(1)} \frac{\partial \Phi^{(1)}}{\partial \xi} + B \frac{\partial^3 \Phi^{(1)}}{\partial \xi^3} = 0 \qquad [5.109]$$

where $A = 1$ is the *coefficient of nonlinearity* and $B = 1/2$ is the *coefficient of dispersion*.

The steady solution of the KdV equation represents a *soliton*, that is a perturbation that propagates in the medium without deforming.

To find the steady stationary solution of the KdV equation, let us introduce the variable:

$$\zeta = \xi - u_0 \tau \qquad [5.110]$$

where $u_0$ is a normalized constant velocity.

The KdV equation [5.109] can then be rewritten as:

$$-u_0 \frac{d\Phi^{(1)}}{d\zeta} + A\Phi^{(1)} \frac{d\Phi^{(1)}}{d\zeta} + B \frac{d^3 \Phi^{(1)}}{d\zeta^3} = 0 \qquad [5.111]$$

This equation can be integrated to have:

$$B \frac{d^2 \Phi^{(1)}}{d\zeta^2} + \frac{A}{2}\Phi^{(1)2} - U_0 \Phi^{(1)} = C \qquad [5.112]$$

where $C$ is an integration constant. By imposing the conditions at the limits appropriate to localized perturbations, namely $\Phi^{(1)} \to 0$, $d\Phi^{(1)}/d\zeta \to 0$, and $d^2\Phi^{(1)}/d\zeta^2 \to 0$, for $\zeta \to \pm\infty$, we have:

$$\Phi(\xi,\tau) = \Phi_m sech^2\left[\frac{(\xi - u_0\tau)}{\Delta}\right] \qquad [5.113]$$

where $sech(x) = 1/\cosh(x)$ is the *hyperbolic secant* function.

In relationship [5.113], $\Phi_m$ and $\Delta$ represent the amplitude and width of the soliton, respectively, and are given by:

$$\Phi_m = \frac{3u_0}{A}$$
$$\Delta = \sqrt{\frac{4B}{u_0}} \qquad [5.114]$$

Figure 5.5 illustrates an example of a soliton (equation [5.113]), for $\Phi_m = \Delta = u_0 = 1$.

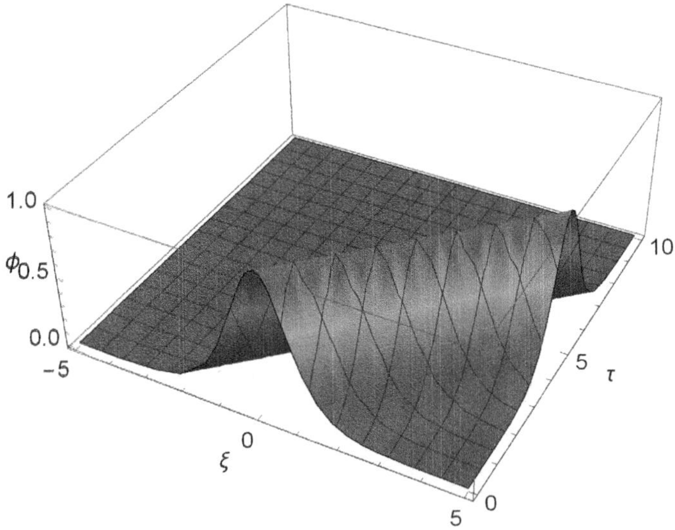


**Figure 5.5.** *Example of a soliton (equation [5.113]), for $\Phi_m = \Delta = u_0 = 1$. For a color version of this figure, see www.iste.co.uk/ourabah/collective.zip*

# 6

# Gravitational Systems

*Even stars collide, and out of their crashing new worlds are born. Today I know "THAT IS LIFE"!*

Charlie Chaplin

This chapter focuses on the application of kinetic and hydrodynamic approaches to self-gravitating particle systems or, in short, gravitational systems. A gravitational system is a system of particles that interact with each other through the forces of gravity. These systems are, indeed, omnipresent in the context of astrophysics and cosmology because dominant interactions, at such scales, are gravitational interactions.

There is a certain analogy between plasmas and gravitational systems. Indeed, in plasmas (assuming ions are at rest), electrons interact with each other essentially through Coulomb interactions, while in a gravitational system, particles interact through the forces of gravity; both decrease with the distance $r$ in $\sim 1/r^2$. The main difference between the two scenarios is that Coulomb interactions between electrons are *repulsive* while gravitational interactions are always *attractive*.

This formal analogy between plasmas and gravitational systems can be used by applying kinetic and hydrodynamic approaches and techniques developed in Chapter 5 in the context of plasmas to the study of gravitational systems. We will see that beyond this formal analogy, oscillations similar to those discussed in the context of plasmas take place in gravitational systems. However, these oscillations have a rather particular nature in that they can become unstable. This is the phenomenon of *gravitational instability* or *Jeans instability* (Jeans 1902).

Physically speaking, Jeans instability occurs when the gravitational pull caused by an increase in density (an overdensity) in a region of the system becomes greater than the pressure forces that tend to relax the overdensity and bring it back to equilibrium density. The Jeans instability mechanism is of paramount importance in astrophysics and cosmology because it is the origin (at least partially) of the formation of all known astrophysical objects, stars, galaxies, galaxy clusters, etc. The mechanism is also of crucial importance in the processes of star formation within giant molecular clouds (see Figure 6.1) and in the initial formation of large structures in the early Universe.

**Figure 6.1.** *Formation of stars in the Circinus molecular cloud. For a color version of this figure, see www.iste.co.uk/ourabah/collective.zip*

COMMENTS ON FIGURE 6.1.– *The colors represent specific wavelengths in the infrared range. Blue and cyan represent wavelengths between 3.4 and 4.6 μm, mostly emanating from stars, while green and red represent wavelengths between 12 and 22 μm, mostly emanating from dust (source: NASA/JPL-Caltech/UCLA).*

## 6.1. Jeans instability: hydrodynamic approach

Firstly, let us study the process of gravitational instability as part of the fluid model. To simplify, let us first consider the total absence of pressure forces (i.e. a zero temperature). Under these conditions, we can take up the hydrodynamic model, in its classic form, described in section 5.1 in the absence of pressure forces, simply by replacing the electric (Coulomb) forces with the force of gravity. We can then express this as:

$$
\begin{aligned}
&\frac{\partial n}{\partial t}+\nabla\cdot(n\mathbf{v})=0\\
&mn\left(\frac{\partial \mathbf{v}}{\partial t}+\mathbf{v}\nabla\mathbf{v}\right)=-mn\nabla\phi
\end{aligned}
\qquad [6.1]
$$

where we preferred to work in terms of the gravitational potential $\phi$ rather than in terms of the gravitational field, $\vec{g}=-\nabla\phi$. System [6.1] must be closed by considering the Poisson equation for the gravitational potential, thus linking potential $\phi$ to density, namely:

$$\nabla^2\phi = 4\pi Gmn \qquad [6.2]$$

By following the same steps described in section 5.1, namely by linearizing the system of equations [6.1] and [6.2], and by searching for a solution in the form of plane waves, we arrive at[1]:

$$\omega^2 = -4\pi Gmn_0 \qquad [6.3]$$

or even:

$$\omega^2 = -\Omega_J^2 \qquad [6.4]$$

where quantity $\Omega_J \equiv \sqrt{4\pi Gmn_0}$ is called the *Jeans frequency*. Comparing relationship [6.4] with its equivalent in the case of a plasma (i.e. equation [5.6]), plasma frequency, playing the role of characteristic frequency or time scale, appears to be replaced here (to a *minus* sign) by the Jeans frequency. However, there is a crucial difference. Indeed, by observing that $\Omega_J^2 \equiv 4\pi Gmn_0 \geq 0$, it is deduced that frequency $\omega$ is necessarily *imaginary*. This tells us that a system where particles interact only through the forces of gravity (in the absence of pressure forces) is *necessarily* unstable.

To better understand this, let us remember that in order to arrive at equation [6.4], we sought solutions in the form of plane waves, the temporal part of which is given by $\sim e^{-i\omega t}$. Now, if $\omega$ is imaginary, we no longer have an oscillatory behavior but two exponential solutions $\sim e^{\pm\Omega_J t}$, corresponding to the two roots of [6.4]. The solution of positive sign $\sim e^{+\Omega_J t}$ is particularly interesting because it indicates that density

1 A small subtlety should be noted here: it should be noted that for a constant gravitational potential $\phi_0$, we have $\nabla^2\phi_0 = 0$ and the Poisson equation ($\nabla^2\phi_0 = 4\pi Gmn_0$) would only be satisfied in this case for a zero density $n_0$, which is of no physical interest. One way to get around this problem is to consider the Poisson equation only for perturbed quantities. This way of circumventing the problem is called the *Jeans Swindle*.

increases exponentially with time and becomes infinite when $t \to \infty$. This shows that a gravitational system is unstable and eventually collapses on itself.

One question that naturally arises is as follows: if gravitational systems are always unstable, how is it that the universe is filled with them? Indeed, at large scales, the Universe is filled with systems interacting essentially through the forces of gravity and these are not always unstable. In fact, this is possible only because there are other repulsive forces (essentially pressure forces) that compensate for the (always attractive) forces of gravity.

Let us go back to the previous model but this time including the effect of *ideal gas* type pressure forces (as in section 5.1). Then, the model becomes:

$$\begin{aligned} &\frac{\partial n}{\partial t} + \nabla \cdot (n\mathbf{v}) = 0 \\ &mn\left(\frac{\partial \mathbf{v}}{\partial t} + \mathbf{v}\nabla\mathbf{v}\right) = -mn\nabla\phi - 3k_B T\nabla n \end{aligned} \quad [6.5]$$

which will also be coupled to the Poisson equation for gravitational potential [6.2]. By following the same steps as before (linearization and looking for solutions in the form of plane waves), we have:

$$\omega^2 = -\Omega_J^2 + \frac{3k_B T}{m}k^2 \quad [6.6]$$

This equation highlights the action of temperature (and therefore pressure forces), which opposes gravitational attraction. In limit $T = 0$, we find relationship [6.4], indicating the instability of gravitational systems. However, when the temperature is high, the second term in equation [6.6] takes over the first term and results in a real $\omega$ frequency, describing simple oscillations as in the case of a plasma.

More quantitatively, we can find the exact limit between a stable regime (oscillations) and an unstable regime (exponential growth over time) from equation [6.6]. This limit corresponds to the value of $k$ for which $\omega^2$ changes signs (thus delimiting the regime where $\omega$ is real of the regime where $\omega$ is imaginary). Putting $\omega^2 = 0$ in equation [6.6], we have:

$$k_J = \sqrt{\frac{4\pi G m^2 n_0}{3k_B T}} \quad [6.7]$$

This quantity is called the *Jeans wavenumber*. For any perturbation whose wavenumber $k$ is greater than $k_J$, there are simple oscillations, whereas for any perturbation whose wavenumber is less than $k_J$, there is instability. It is also possible to define a characteristic wavelength associated with $k_J$, that is, $\lambda_J \equiv 1/k_J$ (*Jeans wavelength*), so that any perturbation whose wavelength exceeds $\lambda_J$ gives rise to instability.

In astrophysics and cosmology, it is more interesting to define the limit between a stable regime (oscillations) and an unstable regime (exponential growth) in terms of mass. We define the *Jeans mass* as being the mass initially contained in a sphere of radius $\lambda_J$, that is:

$$M_J \equiv \frac{4\pi}{3}\lambda_J^3 = \frac{\pi}{6}\sqrt{\frac{1}{mn_0}\left(\frac{\pi k_B T}{Gm}\right)^3} \qquad [6.8]$$

Thus, a gas of particles, of equilibrium density $n_0$ and at temperature $T$, interacting through the forces of gravity, cannot exceed a limit mass, given by equation [6.8]. Beyond this mass, the gas begins to fragment and form smaller mass structures. It is this phenomenon that explains the formation of stars and other structures from the large expanses of gas that populate the Universe.

A kinetic approach to gravitational instability is obviously also possible. It will not be discussed here because it is identical to the kinetic processing of plasma presented in section 5.3 and does not present any additional technical difficulties. However, it is interesting to say a word about the phenomenon of gravitational instability in the quantum case. Thus, if we include the contribution of the Bohm quantum potential in the hydrodynamic model [6.1], we have:

$$\begin{aligned} &\frac{\partial n}{\partial t} + \nabla \cdot (n\mathbf{v}) = 0 \\ &mn\left(\frac{\partial \mathbf{v}}{\partial t} + \mathbf{v}\nabla\mathbf{v}\right) = -mn\nabla\phi + \frac{\hbar^2}{2m}\nabla\left(\frac{\nabla^2\sqrt{n}}{\sqrt{n}}\right) \end{aligned} \qquad [6.9]$$

In the case of plasmas, an approach identical to that presented in section 5.1 gives the following dispersion relation:

$$\omega^2 = -\Omega_J^2 + \frac{\hbar^2 k^4}{4m} \qquad [6.10]$$

Like pressure forces, it indicates that quantum effects also oppose gravitational attraction, allowing gravitational systems to remain stable. This balance between gravity forces and quantum effects plays a particularly important role in some dark matter models, which will be discussed in Chapter 8.

In the same way as in plasmas, pressure forces can be included parallel to the Bohm quantum potential. The result would simply be a combination of dispersion relations [6.6] and [6.10], namely:

$$\omega^2 = -\Omega_J^2 + \frac{3k_BT}{m}k^2 + \frac{\hbar^2 k^4}{4m} \qquad [6.11]$$

indicating that both pressure forces and quantum effects tend to stabilize the gravitational system.

## 6.2. Kinetic approach and collision effect

As mentioned above, we are interested in the hydrodynamic study of gravitational instability in a system assumed to be collision free. In this section, we will present a kinetic analysis of the same problem, taking into account the effect of collisions. A kinetic approach in the absence of collisions would start in the Vlasov equation (or the Boltzmann equation without collisions) and would thus result in dispersion relation [6.6]. To determine the effect of collisions on the mechanism of Jeans instability, it is necessary to go beyond the Vlasov equation and use the Boltzmann equation in a more general form. For the collision term, we will introduce the BGK approximation analyzed in section 1.3.2 of Chapter 1, which remains a fairly good approximation of the collision term. Such a kinetic analysis of Jeans instability was processed for the first time by Trigger et al. (2004). Using the BGK approximation for the collision term, the Boltzmann equation is written as follows:

$$\frac{\partial f}{\partial t} + \mathbf{v}\cdot\nabla f - \nabla\varphi\cdot\frac{\partial f}{\partial \mathbf{v}} = I_{coll} \qquad [6.12]$$

where the collision term is given by:

$$I_{coll} \equiv \left(\frac{\delta f}{\delta t}\right)_{coll} = -\nu\,(f - f_0) \qquad [6.13]$$

$\nu$ is the frequency of collisions. Equation [6.12] will have to be coupled to the Poisson equation for the gravitational potential:

$$\nabla^2\phi = 4\pi Gmn = 4\pi Gm \iiint f d^3v \qquad [6.14]$$

The system of equations thus formed can be processed in a similar way to what has been presented in section 5.3, namely by linearizing equations and by looking for solutions in the form of plane waves. By following these steps, the following integral dispersion relation is easily obtained:

$$\frac{\Omega_J^2}{k^2}\int\frac{\left[\mathbf{k}\cdot\partial f_0(\mathbf{v})/\partial\mathbf{v}+i\nu k^2 f_0(\mathbf{v})/\Omega^2\right]}{\omega+i\nu-\mathbf{k}\cdot\mathbf{v}}d^3\mathbf{v}=1 \qquad [6.15]$$

where $\Omega_J$ is the Jeans frequency. It can already be noted that for $\nu=0$ (in the absence of collisions), we find the integral form of the dispersion relation obtained from the Vlasov equation [5.39].

Suppose that wave vector $\mathbf{k}$ is parallel to the $x$-axis and consider a classic equilibrium distribution given by the Maxwell–Boltzmann distribution:

$$f_0(\mathbf{v})=\frac{1}{(2\pi)^{3/2}v_{th}^3}\exp\left(-\frac{v^2}{2v_{th}^2}\right) \qquad [6.16]$$

In this case, the integral dispersion relation [6.15] can be rewritten in the following form:

$$\frac{\Omega_J^2}{k}\int_{-\infty}^{\infty}\left(\frac{df_0(u)}{du}+i\frac{\nu k f_0(u)}{\Omega_J^2}\right)\frac{du}{(\omega+i\nu-ku)}=1 \qquad [6.17]$$

where $u\equiv v_x$ and:

$$f_0(u)=\int_{-\infty}^{\infty}\int_{-\infty}^{\infty}f_0(\mathbf{v})dv_y dv_z=\frac{1}{\sqrt{2\pi}v_{th}}\exp\left(-\frac{u^2}{2v_{th}^2}\right) \qquad [6.18]$$

represents marginal distribution (integrated on the two remaining components of velocity $v_y$ and $v_z$). Using [6.18], relationship [6.17] can be rewritten as follows:

$$1+\frac{\Omega_J^2}{\sqrt{2\pi}\,(kv_{th})^2}\int_{-\infty}^{\infty}\frac{\exp\left(-x^2/2\right)\left(x-i\nu k v_{th}/\Omega_J^2\right)}{\beta-x}dx=0 \qquad [6.19]$$

where:

$$x=\frac{u}{v_{th}},\quad \beta=\frac{\omega+i\nu}{kv_{th}} \qquad [6.20]$$

It may be useful to introduce the function:

$$J(\beta)=\frac{\beta}{\sqrt{2\pi}}\int_{-\infty}^{\infty}\frac{\exp\left(-x^2/2\right)}{\beta-x}dx \qquad [6.21]$$

and rewrite relationship [6.19] as follows:

$$1-\frac{\Omega_J^2}{(kv_{th})^2}\left[1-J(\beta)+\frac{i\nu\,(kv_{th})^2}{\Omega_J^2(\omega+i\nu)}J(\beta)\right]=0 \qquad [6.22]$$

The usefulness of such a notation is that function $J(\beta)$, thus defined, has a known asymptotic behavior.

Thus, for $|\beta| \ll 1$, we have:

$$J(\beta) \approx -i\sqrt{\frac{\pi}{2}}\beta, \quad \text{with } |\beta| \ll 1 \qquad [6.23]$$

Moreover, for $|\beta| \gg 1$, we have:

$$\begin{aligned} &J(\beta) = -i\sqrt{2\pi}\beta\exp\left(-\beta^2/2\right), \quad \text{with } |\beta| \gg 1 \\ &|\mathrm{Im}(\beta)| \geqslant |\mathrm{Re}(\beta)|, \quad \mathrm{Im}(\beta) < 0 \end{aligned} \qquad [6.24]$$

It is now possible to analyze dispersion relation [6.22] in an approximate manner in order to compare with the collisionless case. Consider the limit:

$$|\omega+i\nu| \geqslant kv_{th}, \quad \text{or } \beta \geqslant 1 \qquad [6.25]$$

and let us take a Taylor expansion of the term $(1-x/\beta)^{-1}$. Taking into account the symmetry (the integral of an odd function over a symmetrical interval is cancelled), we have:

$$1+\frac{\Omega_J^2}{(\omega+i\nu)^2}+\frac{3\,(kv_{th}\Omega_J)^2}{(\omega+i\nu)^4}-\frac{i\nu}{\omega+i\nu}-\frac{i\nu\,(kv_{th})^2}{(\omega+i\nu)^3}=0 \qquad [6.26]$$

Note that for $\nu = 0$ (in the absence of collisions), the above dispersion relation is simply reduced to:

$$\omega^2 = -\Omega_J^2 + 3\,(kv_{th})^2 = -\Omega_J^2 + 3\frac{k_B T}{m}k^2 \qquad [6.27]$$

which is none other than relationship [6.6], obtained from the hydrodynamic model or the Vlasov equation in the collisionless case.

It may be noted that the effect of a non-zero collision frequency $\nu$ appears in dispersion relation [6.26] in an imaginary form. This means that collisions do not affect the instability criterion (Jeans wavenumber $k_J$ or Jeans wavelength $\lambda_J$ remain unchanged). In contrast, collisions alter the growth rate of the perturbation. This is particularly easy to see in the case of long wavelengths ($k \to 0$). Within this limit, we obtain from the dispersion relation [6.26] (for an unstable mode, i.e. $\mathrm{Im}(\omega) > 0$):

$$\omega_0 \equiv \omega(k=0) = \frac{i}{2}\left(\sqrt{\nu^2 + 4\Omega_J^2} - \nu\right) \qquad [6.28]$$

This latter relationship shows that collisions have the effect of decreasing the imaginary part of the frequency, thus slowing down the process of gravitational instability (see Figure 6.2, showing the imaginary part of the frequency [6.28]).

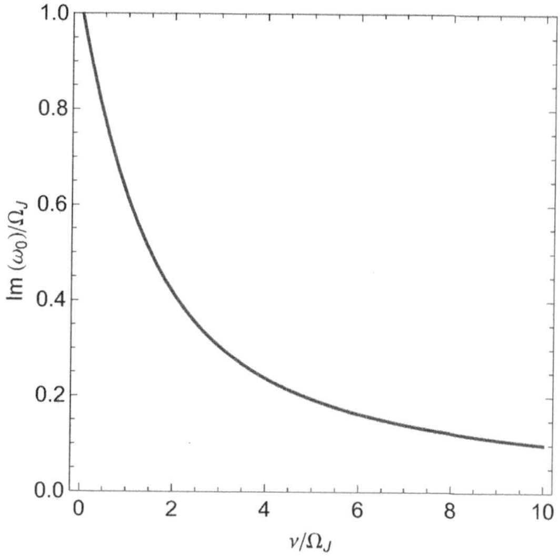


**Figure 6.2.** *Imaginary part of $\omega_0$ (given by relationship [6.28]), as a function of collision frequency $\nu$, both normalized to $\Omega_J$*

## 6.3. Jeans instability in the presence of dark matter

The results discussed above can easily be extended to the case involving several species of particles. An interesting situation from a phenomenological point of view is that of a gravitational system composed of ordinary (baryonic) matter and dark matter (the usefulness of kinetic and hydrodynamic approaches in the study of dark matter will be discussed in more detail in Chapter 8). Intuitively, it may seem natural to think that the presence of dark matter reduces the stability of gravitational systems, due to the gravitation created by dark matter particles. Also, it is possible to study this qualitatively both in the kinetic approach and in the hydrodynamic approach. Here, we will present two distinct approaches to the mechanism of gravitational instability in the presence of dark matter: one kinetic (Kremer and André 2016) and the other hydrodynamic (Tsiklauri 1998).

### 6.3.1. *Kinetic approach*

Let us start by looking at the problem in the kinetic context. In what follows, we consider the absence of collisions: thus, it will be necessary to write collisionless Boltzmann equations (Vlasov equations) for the two species (baryonic matter and dark matter), namely:

$$\begin{aligned} &\frac{\partial f_b}{\partial t} + \mathbf{v}_b \cdot \frac{\partial f_b}{\partial \mathbf{x}} - \nabla\Phi \cdot \frac{\partial f_b}{\partial \mathbf{v}_b} = 0 \\ &\frac{\partial f_d}{\partial t} + \mathbf{v}_d \cdot \frac{\partial f_d}{\partial \mathbf{x}} - \nabla\Phi \cdot \frac{\partial f_d}{\partial \mathbf{v}_d} = 0 \end{aligned} \qquad [6.29]$$

where index $b$ refers to baryonic matter while index $d$ (for *dark matter*) refers to dark matter. Kinetic equations [6.29] do not constitute a closed system of equations and must be coupled to the Poisson equation of gravitational potential (in what follows, we will ignore corrections due to general relativity), namely[2]:

$$\nabla^2\Phi = 4\pi G\left(\iiint f_b d\mathbf{v}_b + \iiint f_d d\mathbf{v}_d\right) = 4\pi G(\rho_b + \rho_d) \qquad [6.31]$$

Equations [6.29] and [6.31] can be processed as in sections 5.3 and 6.2, by linearizing them around equilibrium values of $f_b$, $f_d$, and $\Phi$, i.e. $f_b^0$, $f_d^0$ and $\Phi_0$ and by searching for solutions in the form of plane waves, which amounts to writing:

$$\begin{aligned} &f(\mathbf{x},\mathbf{v}_b,t) = f_b^0(\mathbf{v}_b) + \bar{f}_b^1 \exp[i(\mathbf{k}\cdot\mathbf{x} - \omega t)] \\ &f(\mathbf{x},\mathbf{v}_d,t) = f_d^0(\mathbf{v}_d) + \bar{f}_d^1 \exp[i(\mathbf{k}\cdot\mathbf{x} - \omega t)] \\ &\Phi(\mathbf{x},t) = \Phi_0(\mathbf{x}) + \bar{\Phi}_1 \exp[i(\mathbf{k}\cdot\mathbf{x} - \omega t)] \end{aligned} \qquad [6.32]$$

2 To simplify notations, distributions are normalized here to mass density $\rho = mn$, namely:

$$\iiint f_{b,d} d\mathbf{v}_{b,d} = \rho_{b,d} \qquad [6.30]$$

where quantities $\bar{f}_b^1$, $\bar{f}_d^1$, and $\bar{\Phi}_1$ represent small perturbation amplitudes. By replacing quantities [6.32] in equations [6.29] and [6.31], and keeping only the terms of the first order, we have:

$$
\begin{aligned}
&-i\omega \bar{f}_b^1 + \mathbf{v}_b \cdot \left(i\mathbf{k}\bar{f}_b^1\right) - \left(i\mathbf{k}\bar{\Phi}_1\right) \cdot \frac{\partial f_b^0}{\partial \mathbf{v}_b} = 0 \\
&-i\omega \bar{f}_d^1 + \mathbf{v}_d \cdot \left(i\mathbf{k}\bar{f}_d^1\right) - \left(i\mathbf{k}\bar{\Phi}_1\right) \cdot \frac{\partial f_d^0}{\partial \mathbf{v}_d} = 0 \\
&-\mathbf{k}^2\bar{\Phi}_1 = 4\pi G\left(\iiint \bar{f}_b^1 d\mathbf{v}_b + \iiint \bar{f}_d^1 d\mathbf{v}_d\right)
\end{aligned}
\qquad [6.33]
$$

where $\mathrm{k} = |\mathbf{k}|$ is the modulus of the wavenumber. By combining the above three equations, the following integral dispersion relation is obtained:

$$
\mathrm{k}^2 + 4\pi G\mathbf{k} \cdot \left(\iiint \frac{\partial f_b^0}{\partial \mathbf{v}_b} \frac{d\mathbf{v}_b}{\mathbf{v}_b \cdot \mathbf{k} - \omega} + \iiint \frac{\partial f_d^0}{\partial \mathbf{v}_d} \frac{d\mathbf{v}_d}{\mathbf{v}_d \cdot \mathbf{k} - \omega}\right) = 0 \qquad [6.34]
$$

Let us explicitly study the case where equilibrium distributions $f_b^0$ and $f_d^0$ are given by the Maxwell–Boltzmann distribution:

$$
\begin{aligned}
f_b^0(\mathbf{v}_b) &= \frac{\rho_b^0}{\left(2\pi\sigma_b^2\right)^{\frac{3}{2}}} e^{-v_b^2/2\sigma_b^2} \\
f_d^0(\mathbf{v}_d) &= \frac{\rho_d^0}{\left(2\pi\sigma_d^2\right)^{\frac{3}{2}}} e^{-v_d^2/2\sigma_d^2}
\end{aligned}
\qquad [6.35]
$$

This amounts to conventionally processing baryonic matter and dark matter and assuming that they are both initially in equilibrium. Quantities $\rho_b^0$ and $\rho_d^0$ represent the equilibrium mass densities of baryonic matter and dark matter, respectively, while quantities $\sigma_b$ and $\sigma_d$ represent thermal velocities of each species. Let us consider, without a loss of generality, that the wave vector is parallel to the $x$-axis. Using equilibrium distributions [6.35] and integral dispersion relation [6.34], we can write:

$$
\mathrm{k}^2 = 4\pi G \frac{2}{\sqrt{\pi}} \left[\frac{\rho_b^0}{\sigma_b^2} \int_0^\infty \frac{x^2 e^{-x^2}}{x^2 - w^2/\left(2\sigma_b^2 \mathrm{k}^2\right)} + \frac{\rho_d^0}{\sigma_d^2} \int_0^\infty \frac{y^2 e^{-y^2}}{y^2 - w^2/\left(2\sigma_d^2 \mathrm{k}^2\right)}\right] \qquad [6.36]
$$

where dimensionless quantities:

$$
x = \frac{v_{bx}}{\sqrt{2}\sigma_b}, \quad y = \frac{v_{dx}}{\sqrt{2}\sigma_d} \qquad [6.37]
$$

have been introduced. These simply represent velocities of each species, normalized to the corresponding thermal velocity.

As discussed previously, instability appears when perturbations no longer correspond to oscillations but to an exponential growth with time, that is for $\Re(\omega) = 0$ and $\Im(\omega) > 0$. Denoting $\omega = i\omega_i$ ($\omega_i$ being the imaginary part of $\omega$), we have the identity:

$$\int_0^\infty \frac{x^2 e^{-\mu^2 x^2}}{x^2+\beta^2} dx = \frac{\sqrt{\pi}}{2\mu} - \frac{\pi\beta}{2} e^{\beta^2} \operatorname{erfc}(\beta\mu) \tag{6.38}$$

where:

$$\operatorname{erfc}(x) = \frac{2}{\sqrt{\pi}} \int_x^\infty e^{-t^2}\, dt \tag{6.39}$$

is the *complementary error function*. Using identity [6.38] and equation [6.34], we have (for $\omega = i\omega_i$):

$$\begin{aligned} k_*^2 = 1 &- \sqrt{\frac{\pi}{2}} \frac{\omega_*}{k_*} \exp\left(\frac{\omega_*^2}{2k_*^2}\right) \operatorname{erfc}\left(\frac{\omega_*}{\sqrt{2}k_*}\right) \\ &+ \frac{\rho_b^0 \sigma_d^2}{\rho_d^0 \sigma_b^2} \left[1 - \sqrt{\frac{\pi}{2}} \frac{\sigma_d}{\sigma_b} \frac{\omega_*}{k_*} \exp\left(\frac{\sigma_d^2}{\sigma_b^2} \frac{\omega_*^2}{2k_*^2}\right) \operatorname{erfc}\left(\frac{\sigma_d}{\sigma_b} \frac{\omega_*}{\sqrt{2}k_*}\right)\right] \end{aligned} \tag{6.40}$$

where:

$$\mathrm{k}_* = \frac{\sigma_d \mathrm{k}}{\sqrt{4\pi G \rho_d^0}} \tag{6.41}$$

and:

$$\omega_* = \frac{\omega_i}{\sqrt{4\pi G \rho_d^0}} \tag{6.42}$$

represent the dimensionless wavenumber and the dimensionless frequency, respectively.

Equation [6.40] makes it possible to analyze gravitational instability and to determine the zones of instability in plane $(\omega_*, k_*)$. Equation [6.40] uses density ratio $\rho_d^0/\rho_b^0$ and thermal velocities $\sigma_d/\sigma_b$. Thus, to be able to draw quantitative conclusions from equation [6.40], we must specify the numerical value of these ratios. For $\rho_d^0/\rho_b^0$, as this ratio has not changed much during the evolution of the Universe, it is possible to use the current density ratio of baryonic matter and dark matter (Kremer and André 2016), namely, $\rho_d^0/\rho_b^0 \approx 5.5$. For the ratio of thermal velocities $\sigma_d/\sigma_b$, simulations on the "Milky Way" type galaxies point toward an approximate value $\sigma_d/\sigma_b = 170/93 \approx 1{,}83$ (Ling et al. 2010). With this, we can quantitatively discuss gravitational instability in the presence of dark and baryonic matter.

Considering $\omega_* = 0$ in equation [6.40], it is possible to determine the numerical value of dimensionless wavenumber $k_* = 1.2694$. This value can be interpreted as the ratio of the Jeans number of the system, composed of baryonic matter and dark matter $\mathrm{k}_J^{db}$, and the Jeans number for dark matter only, that is, $\mathrm{k}_J^d = \sqrt{4\pi G\rho_d^0}/\sigma_d$. In the same way, it is possible to determine the ratio of the Jeans wavelengths $\lambda_J^{db} = 2\pi/\mathrm{k}_J^{db}$ and $\lambda_J^d = 2\pi/\mathrm{k}_J^d$ which read as $\lambda_J^{db}/\lambda_J^d = k_J^d/_J^{db} = 0.7877$. This ratio, being less than 1, indicates that the Jeans wavelength for a system composed of dark matter and baryonic is *lower* than that of a system composed only of dark matter. Similarly, from the ratio of wavelengths, it is possible to determine the ratio between Jeans masses as follows:

$$\frac{M_J^{db}}{M_J^d} = \frac{\rho_d^0 + \rho_b^0}{\rho_d^0}\left(\frac{\lambda_J^{db}}{\lambda_J^d}\right)^3 = \frac{\rho_d^0 + \rho_b^0}{\rho_d^0}\sqrt{\left(\frac{\rho_d^0\sigma_b^2}{\rho_d^0\sigma_b^2 + \rho_b^0\sigma_d^2}\right)^3} = 0.5791 \qquad [6.43]$$

This indicates that in the presence of dark matter, a lower Jeans mass is sufficient to initiate gravitational instability and the formation of structures.

To clearly see the role of the ratio of thermal velocities $\sigma_d/\sigma_b$, Figure 6.3 shows plane $(\omega_*^2, k_*^2)$ for different values of ratio $\sigma_d/\sigma_b$. Regions below the lines refer to the positive values of the outline of equation [6.40], while regions above the lines refer to negative values. The arrow indicates the increasing direction of ratios $\sigma_d/\sigma_b$ considered. It can be seen, here again, that by increasing ratio $\sigma_d/\sigma_b$, the zone of instability increases and, consequently, the Jeans mass decreases. Thus, for example $M_J^{db}/M_J^d = 0.8338$ for $\sigma_d/\sigma_b = 1.2$ and $M_J^{db}/M_J^d = 0.4585$ for $\sigma_d/\sigma_b = 2.2$ (see Table 6.1 for numerical values of $M_J^{db}/M_J^d$ for different values of $\sigma_d/\sigma_b$).

| $\sigma_d/\sigma_b$ | 1.00 | 1.20 | 1.40 | 1.60 | 1.83 | 2.00 | 2.20 |
|---|---|---|---|---|---|---|---|
| $M_J^{\mathrm{db}}/M_J^{\mathrm{d}}$ | 0.9199 | 0.8338 | 0.7481 | 0.6662 | 0.5791 | 0.5206 | 0.4585 |

**Table 6.1.** *Values of $M_J^{db}/M_J^d$ for different values of ratio $\sigma_d/\sigma_b$, for $\rho_d^0/\rho_b^0 = 5.5$*

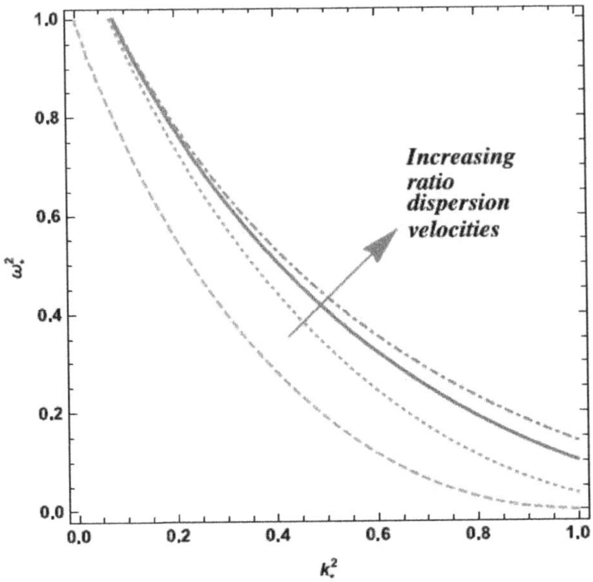


**Figure 6.3.** *Relationship between the dimensionless wavenumber and frequency, for different values of $\sigma_d/\sigma_b$: $\sigma_d/\sigma_b = 1.83$ (solid line), $\sigma_d/\sigma_b = 1.20$ (dotted line) and $\sigma_d/\sigma_b = 2.20$ (dash-dotted line). The dotted curve represents the case of dark matter only, that is in the absence of baryonic matter (source: Kremer and André (2016)). For a color version of this figure, see www.iste.co.uk/ourabah/collective.zip*

### 6.3.2. *Hydrodynamic approach*

In the same way, Jeans instability in the presence of dark matter and baryonic can be processed within the hydrodynamic model context. We will present here the results of the Tsiklauri hydrodynamic model (Tsiklauri 1998).

Since this is a problem involving two species (dark matter and baryonic), the model will be formed from fluid equations of each species, coupled with the Poisson equation for gravitational potential. As we have seen in Chapter 2, hydrodynamic equations do not form a closed system and must be supplemented with an equation of state, linking pressure to density. The Tsiklauri hydrodynamic model is represented by the following system of equations:

$$\begin{aligned}
&\frac{\partial \rho_i}{\partial t} + \nabla(\rho_i V_i) = 0 \\
&\frac{\partial V_i}{\partial t} + (V_i \cdot \nabla) V_i = -\nabla \varphi_i - \frac{\nabla P_i}{\rho_i} \\
&P_i = K_i \rho_i^{\gamma_i} \\
&\Delta \varphi_i = 4\pi G \sum_i \rho_i
\end{aligned} \qquad [6.44]$$

where $\rho_i$, $v_i$, $P_i$ and $\varphi_i$ represent the density, velocity, pressure and gravitational potential, respectively, of each species. Index $i = g, w$ refers to the species in question, index $g$ refers to interstellar gas (baryonic or ordinary matter) and index $w$ refers to a particular model of dark matter, called *Weakly Interacting Massive Particles* (WIMP).

Note that here, contrary to what has been done in kinetic processing (section 6.3.1), the model consists of considering two contributions to gravitational potential (one for each species) because this model involves WIMP-type dark matter, supposed to interact very weakly with ordinary matter.

By following the usual steps, that is, by separating each quantity ($f$) into an equilibrium quantity and a perturbation that is assumed to be small:

$$f = f^0 + f' \exp[i(kr - \omega t)] \qquad [6.45]$$

and thus linearizing the system of equations [6.44], we arrive at the following dispersion relation:

$$\omega^2 = \frac{1}{2}\left\{ \left(\omega_g^J\right)^2 + \left(\omega_w^J\right)^2 \pm \sqrt{\left[\left(\omega_\beta^J\right)^2 - (\omega_w^J)^2\right]^2 + 4\delta^2} \right\} \qquad [6.46]$$

where:

$$\begin{aligned}
&\left(\omega_\theta^J\right)^2 \equiv 4\pi G \rho_g^0 - c_g^2 k^2 \\
&\left(\omega_w^J\right)^2 \equiv 4\pi G \rho_w^0 - c_w^2 k^2
\end{aligned} \qquad [6.47]$$

and expressions:

$$\begin{aligned}
&c_g \equiv \left(\gamma_g P_g^0 / \rho_g^0\right)^{1/2} \\
&c_w \equiv \left(\gamma_w P_w^0 / \rho_w^0\right)^{1/2}
\end{aligned} \qquad [6.48]$$

represent the speed of sound in each of the two fluids (interstellar gas and WIMP-type dark matter). Quantity $\delta \equiv 4\pi G \left(\rho_g^0 \rho_w^0\right)^{1/2}$, on the other hand, represents the coupling between the two species.

From dispersion relation [6.46], the instability condition can be obtained by assuming $\omega^2 = 0$. Indeed, for $\omega^2 > 0$, frequency is real and perturbations behave like oscillations, whereas for $\omega^2 < 0$, frequency becomes imaginary and perturbations increase exponentially with time. The following criterion is thus obtained:

$$\left(\omega_g^J\right)^2 \left(\omega_w^J\right)^2 < \delta^2 \qquad [6.49]$$

By introducing the Jeans wavelength for gas and dark matter fluid, in the usual way:

$$\lambda_g^J = c_\theta \sqrt{\frac{\pi}{G\rho_\theta^0}} \lambda_w^J = c_w \sqrt{\frac{\pi}{G\rho_w^0}} \qquad [6.50]$$

the instability criterion can be written, in terms of wavelength, as follows:

$$\lambda > \frac{\lambda_g^J}{\sqrt{1 + \left(\lambda_g^J / \lambda_w^J\right)^2}} \equiv \lambda^J \qquad [6.51]$$

Thus, if only gas was considered and dark matter was not taken into account, the criterion would be reduced to $\lambda > \lambda_g^J$. The presence of dark matter modifies this criterion by a factor of $1/\sqrt{\left[1 + \left(\lambda_g^J/\lambda_w^J\right)^2\right]}$. The consequence is that the Jeans wavelength is reduced. The presence of dark matter then tends to make the gas more gravitationally unstable, in accordance with the result established in the kinetic context (section 6.3.1).

## 6.4. Jeans instability: alternative theories

Several variants of the Jeans instability study, both in the hydrodynamic model and in kinetic theory, can be discussed, considering previously unknown effects. It may be interesting to extend this study to the context of alternative theories of gravitation. These alternative theories attempt to alleviate the problem of dark matter and dark energy in a phenomenological way. Indeed, dark matter remains a hypothetical component to this day and no observation revealing the nature of dark matter could be made at present. These alternative theories propose modifying the laws of gravity so as not to resort to the introduction of these exotic components that are dark matter and dark energy.

It is interesting to see how fluid and kinetic methods remain valid in such alternative theories and how they allow for exploring observational consequences of these theories. Only one theory in particular will be considered here, namely *Eddington-inspired Born-Infeld gravity* (EiBI) (Pani et al. 2012) and we will present the kinetic study of Jeans instability within the context of this theory (De Martino and Capolupo 2017). A similar approach can be applied in the hydrodynamic context. Furthermore, we will consider only the Newtonian limit of this theory, that is, without taking into account the effects of general relativity.

In the Newtonian limit (weak gravitational fields), EiBI gravitation leads to the following modified Poisson equation (De Martino and Capolupo 2017):

$$\nabla^2\Phi(\mathbf{r},t) = 4\pi G\rho(\mathbf{r},t) + \frac{\kappa}{4}\nabla^2\rho(\mathbf{r},t) \qquad [6.52]$$

where parameter $\kappa$ is a dimensionless real number: this is a free parameter of the theory. In the case $\kappa = 0$, the previous equation is reduced to the ordinary Poisson equation for the gravitational potential and the EiBI gravitation is reduced to the classic theory (general relativity or Newtonian gravitation).

To look at the problem kinetically, equation [6.52] will be coupled to the Vlasov equation (for simplicity, the effect of collisions will be ignored). Then, the system of equations to be considered is as follows:

$$\begin{aligned}
&\left[\frac{\partial}{\partial t} + (\mathbf{v}\cdot\nabla) - \left(\nabla\Phi(\mathbf{r},t)\cdot\frac{\partial}{\partial\mathbf{v}}\right)\right] f(\mathbf{r},\mathbf{v},t) = 0 \\
&\nabla^2\Phi(\mathbf{r},t) = \left[4\pi G + \frac{\kappa}{4}\nabla^2\right]\iiint f(\mathbf{r},\mathbf{v},t)d\mathbf{v}
\end{aligned} \qquad [6.53]$$

We will proceed in a standard way by linearizing the system of equations [6.52] by considering small perturbations around equilibrium values:

$$\begin{aligned}
&f(\mathbf{r},\mathbf{v},t) = f_0(\mathbf{r},\mathbf{v}) + \bar{f}_1\exp[i(\mathbf{k}\cdot\mathbf{x} - \omega t)] \\
&\Phi(\mathbf{r},t) = \Phi_0(\mathbf{r}) + \bar{\Phi}_1\exp[i(\mathbf{k}\cdot\mathbf{x} - \omega t)]
\end{aligned} \qquad [6.54]$$

After linearization, the system of equations [6.53] becomes:

$$\begin{aligned}
&-i\omega f_1 + \mathbf{v}\cdot(i\mathbf{k}f_1) - (i\mathbf{k}\Phi_1)\cdot\frac{\partial f_0}{\partial\mathbf{v}} = 0 \\
&-k^2\Phi_1 = 4\pi G\iiint f_1(\mathbf{r},\mathbf{v},t)\mathrm{d}\mathbf{v} - \frac{\kappa}{4}k^2\iiint f_1(\mathbf{r},\mathbf{v},t)\mathrm{d}\mathbf{v}
\end{aligned} \qquad [6.55]$$

Combining the above two equations, the dispersion relation can be written in the following integral form:

$$1+\left(\frac{4\pi G}{k^2}-\frac{\kappa}{4}\right)\iiint\frac{\mathbf{k}\cdot\frac{\partial f_0}{\partial\mathbf{v}}}{\mathbf{v}\cdot\mathbf{k}-\omega}\mathrm{d}\mathbf{v}=0 \qquad [6.56]$$

Considering the case of a classic system whose equilibrium distribution $f_0$ is given by the Maxwell–Boltzmann distribution [6.35], relationship [6.56] can be written as:

$$1-\left(\frac{4\pi G}{k^2}-\frac{\kappa}{4}\right)\frac{\rho_0}{\sqrt{2\pi\sigma^3}}k\int\frac{ue^{-\frac{u^2}{2\sigma^2}}}{ku-\omega}du=0 \qquad [6.57]$$

where, without losing in generality, wave vector **k** has been assumed to be parallel to the $x$-axis and we have noted that $u\equiv v_x$.

From the above relationship, the criterion of instability in the context of EiBI gravitation can be determined. As discussed above, instability occurs when $\omega$ becomes imaginary or when $\omega^2$ becomes negative. By equating $\omega=0$ in relationship [6.57], the limit wavenumber that separates an oscillatory regime and an instability regime can be determined. We get this from:

$$k^{*2}(\omega=0)=\left(\frac{\sigma^2}{4\pi G\rho_0}+\frac{\kappa}{16\pi G}\right)^{-1}=\left(1+\frac{\kappa}{16\pi G}k_J^2\right)^{-1}k_J^2 \qquad [6.58]$$

where:

$$k_J=\sqrt{\frac{4\pi G\rho_0}{\sigma^2}} \qquad [6.59]$$

is the usual Jeans number [6.7], obtained by assuming $\kappa=0$.

From $k^*$ (equation [6.58]), the modified Jeans wavelength $\lambda^*\equiv 2\pi/k^*$ can be deduced, that is:

$$\lambda^*=\left(1+\frac{\kappa}{16\pi G}k_J^2\right)\lambda_J \qquad [6.60]$$

where $\lambda_J=2\pi/k_J$. It is also possible to determine the Jeans mass, defined as the mass initially contained in a sphere of radius $\lambda^*$, as follows:

$$M^*=\frac{\pi}{6}\sqrt{\frac{1}{\rho_0}\left(\frac{\pi\sigma^2}{G}+\frac{4\pi^2\rho_0\kappa}{16\pi G}\right)^3}=\left(1+k_J^2\frac{\kappa}{16\pi G}\right)^{3/2}M_J \qquad [6.61]$$

where $M_J$ is the usual Jeans mass (equation [6.8]), formed from $\lambda_J$. It can thus be seen that for $\kappa > 0$, mass $M^*$ is greater than $M_J$ and the system remains stable at higher masses. On the contrary, for $\kappa < 0$, the system is unstable for weaker masses. The usual results, within the Newtonian limit, are found in the case of $\kappa = 0$ ($M^* = M_J$ in this case).

Note that the previous analysis can easily be extended to the quantum case, using the quantum hydrodynamic model or the quantum Wigner–Moyal kinetic approach. In this case, we get the dispersion relation from (Bessiri et al. 2021):

$$\omega^2 = -\Omega_J^2 + \frac{\hbar^2}{4m^2}k^4 + \frac{\kappa}{4}\rho_0 k^2 \qquad [6.62]$$

Thus, we find the effect produced by quantum pressure (see equation [6.10]), which opposes gravitational instability. From dispersion relation [6.62], we can find the critical wavenumber demarcating between the oscillatory regime and the unstable regime as follows:

$$\mathrm{k}^* = k_J \left( \sqrt{1 + \left( \frac{\kappa m \rho_0}{4\Omega_J \hbar} \right)^2} - \frac{\kappa m \rho_0}{4\Omega_J \hbar} \right)^{1/2} \qquad [6.63]$$

Previously drawn conclusions remain valid in this case: the system is more stable for $\kappa > 0$ and less stable for $\kappa < 0$.

Before concluding this section, it may be interesting to note that the mass of astrophysical objects can be measured by spectroscopic methods. The same applies to density $\rho_0$ or thermal velocity $\sigma$. Thus, the numerical value of the Jeans mass can be determined from these measurements. Moreover, it is possible to know whether these systems are stable or unstable by observing whether the phenomenon of star formation takes place there. The purpose of this kind of exercise in calculating the Jeans mass in alternative theories is, therefore, to compare the predictions of these theories with observations. Indeed, in addition to EiBI gravitation, the consequences of other alternative theories (such as $f(R)$ theories and theories with non-minimal coupling between matter and curvature) on Jeans instability have been extensively studied in recent literature (Capoziello et al. 2012; Arbuzova et al. 2014; Gomes 2020).

# 7

# Bose-Einstein Condensates

*Knowledge is acquired*
*through experience,*
*everything else is information.*

Albert Einstein

This chapter will focus on the application of kinetic and hydrodynamic approaches to Bose–Einstein condensates. A Bose–Einstein condensate is a *state of matter*, formed of identical bosons (usually atoms behaving like bosons), so that at a sufficiently low temperature, many of the particles making up the system occupy the lowest energy state (ground state). This is only possible for bosons (particles with an integer spin) unlike fermions (particles with a half-integer spin) due to the Pauli exclusion principle.

The prediction of Bose–Einstein condensates dates back to 1924 with the work of *Albert Einstein* and the Indian physicist *Satyendranath Bose*. Einstein predicted that for a boson gas, there is a critical temperature below which a *macroscopic* number of particles making up the system occupy the lowest energy state, thus forming a condensate.

The first experimental observation of the condensation of an atomic gas took place in 1995 because of the work of *Eric Cornell* and *Carl Wieman*, who managed to cool a rubidium gas to temperatures below 170 nK. This experiment earned them the 2001 Nobel Prize in Physics (shared with *Wolfgang Ketterle*). Since then, Bose–Einstein condensates are regularly produced in laboratories. Due to the very low temperature, atoms forming the condensate can be easily controlled (see Figures 7.1 and 7.2), paving the way for new experimental perspectives. Bose–Einstein condensates also find technological and industrial applications, for example, in the production of atomic lasers.

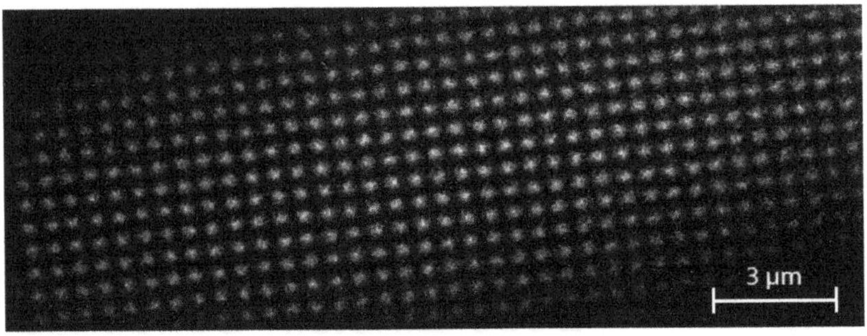


**Figure 7.1.** *Image of a Bose–Einstein condensate in a two-dimensional optical network obtained with an electron microscope (source: Würtz et al. (2009)). For a color version of this figure, see www.iste.co.uk/ourabah/collective.zip*

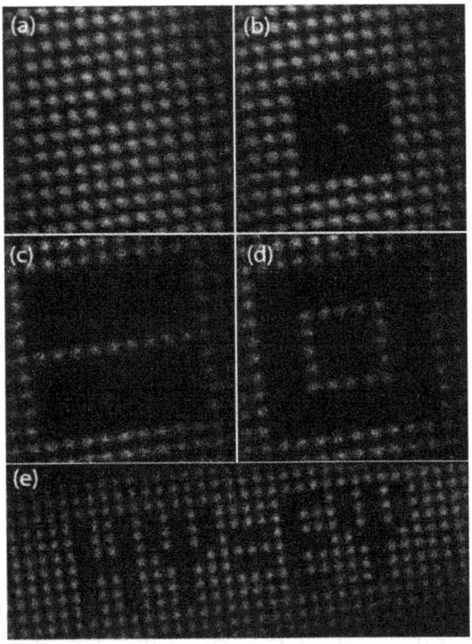


**Figure 7.2.** *Example of manipulations of a Bose–Einstein condensate in a two-dimensional optical network, with a spacing of 600 nm between atoms (source: Würtz et al. (2009)). For a color version of this figure, see www.iste.co.uk/ourabah/collective.zip*

As in a Bose–Einstein condensate, a macroscopic number of quantum particles occupies the same energy level, particles lose their individuality and the condensate must be described as a whole as a macroscopic quantum object. In particular, the wave-like aspect of matter (see Chapter 3) becomes visible on a large scale. Thus, a Bose–Einstein condensate must necessarily be processed in the quantum context. In this chapter, we will see precisely how the quantum hydrodynamic and kinetic models, discussed in Chapter 3, can be adapted to describe condensates. We will also see that there is a phenomenon analogous to plasma oscillations that takes place in condensates: *Bogoliubov oscillations*.

In general, Bose–Einstein condensates are described by a nonlinear form of the Schrödinger equation, called the *Gross–Pitaevskii equation*, which is given by:

$$i\hbar\frac{\partial\Psi(\mathbf{r},t)}{\partial t}=\left[-\frac{\hbar^2}{2m}\nabla^2+U(\mathbf{r})+g|\Psi(\mathbf{r},t)|^2\right]\Psi(\mathbf{r},t) \qquad [7.1]$$

This differs from the (linear) Schrödinger equation only by the appearance of a nonlinear term (proportional to $g$). In equation [7.1], $U$ is an external potential while the nonlinear term, proportional to $g$, describes the interaction potential between bosons ($g$ being a constant depending on the diffusion length of the interaction potential). In what follows, we will see the hydrodynamic and kinetic representations of the Gross–Pitaevskii equation [7.1] and how these representations can be used to describe collective phenomena in condensates.

## 7.1. Quantum hydrodynamic representation

Let us start by discussing the hydrodynamic formulation of the Gross–Pitaevskii equation [7.1]. Going along with what has been done in section 3.2, we introduce the Madelung transformation, namely:

$$\Psi=\sqrt{n}e^{iS/\hbar},\quad \mathbf{v}=\frac{\nabla S}{m} \qquad [7.2]$$

where $n=|\Psi|^2$ represents density and $\mathbf{v}$ the fluid velocity. By introducing transformation [8.32] into the Gross–Pitaevskii equation [7.1] and separating the real part from the imaginary part, the imaginary part gives the continuity equation:

$$\frac{\partial n}{\partial t}+\nabla(n\mathbf{v})=0 \qquad [7.3]$$

which is not affected by the nonlinear term. The real part, on the other hand, gives the following Euler type equation:

$$\frac{\partial \mathbf{v}}{\partial t} + \mathbf{v}\cdot\nabla\mathbf{v} = \nabla U + \frac{\hbar^2}{2m^2}\nabla\left(\frac{\nabla^2\sqrt{n}}{\sqrt{n}}\right) - g\nabla n \qquad [7.4]$$

where we can see that the nonlinearity of the Gross–Pitaevskii equation (term proportional to $g$) produces a term formally identical to an isothermal pressure (see equation [2.37] in Chapter 2). The system formed by equations [7.3]–[7.4] allows for a hydrodynamic description of the Bose–Einstein condensates and the same perturbative techniques, presented in Chapters 5 and 6, can be used in this case.

Here, we present a simple example, which consists of calculating the speed of sound in a condensate. As in the previous chapters, by breaking down quantities into an equilibrium component and a perturbed component:

$$\begin{aligned} n(\mathbf{r},t) &= n_0(\mathbf{r}) + \delta n(\mathbf{r},t) \\ \mathbf{v}(\mathbf{r},t) &= \mathbf{0} + \delta\mathbf{v}(\mathbf{r},t) \\ U(\mathbf{v},t) &= U_0(\mathbf{r}) + \delta U(\mathbf{r},t) \end{aligned} \qquad [7.5]$$

and by linearizing system [7.3]–[7.4], it is easily shown that the perturbations of density in a Bose–Einstein condensate are described by the following relationship:

$$\frac{\partial^2 \delta n}{\partial t^2} - \nabla\left[\frac{g n_0(\mathbf{r})}{m}\nabla\delta n\right] = 0 \qquad [7.6]$$

By seeking solutions in the form of plane waves $\delta n \sim \exp[i(\mathbf{k}\cdot\mathbf{r} - \omega t)]$, we get the following dispersion relation:

$$\omega^2 = c_s^2 k^2 \qquad [7.7]$$

where:

$$c_s = \sqrt{g n_0/m} \qquad [7.8]$$

This simply corresponds to the speed of sound in the condensate (see the speed of sound in neutral gases, discussed in section 5.2). In the case of sound waves propagating in a Bose–Einstein condensate, these are generally referred to as Bogoliubov oscillations.

## 7.2. Quantum kinetic representation

Let us now turn to the kinetic representation of the Gross–Pitaevskii equation [7.1]. For this, we will follow the same steps described in section 3.3 of Chapter 3 for the Schrödinger equation. The only difference in the present case is the presence of the nonlinear term $g|\Psi(\mathbf{r},t)|^2$.

Let us consider two pairs of space and time coordinates, $(\mathbf{r}_1,t_1)$ and $(\mathbf{r}_2,t_2)$. Using notation $\psi_j \equiv \psi(\mathbf{r}_j,t_j)$ and $V_j \equiv V(\vec{r}_j,t_j)$, we can thus rewrite the Gross–Pitaevskii equation [7.1] as follows:

$$\left(\frac{\hbar^2}{2m}\nabla_j^2 - i\hbar\frac{\partial}{\partial t_j}\right)\psi_j = -V_j\psi_j \qquad [7.9]$$

where $V = U + g|\Psi(\vec{r},t)|^2$. By multiplying equation [7.9] for $j=1$ by $\psi_2^*$ and the complex conjugate of the equation for $j=2$ by $\psi_1$ and distinguishing between them, we have:

$$\left[\frac{\hbar^2}{2m}\left(\nabla_1^2 - \nabla_2^2\right) - i\hbar\left(\frac{\partial}{\partial t_1} + \frac{\partial}{\partial t_2}\right)\right]C_{12} = -(V_1 - V_2)C_{12} \qquad [7.10]$$

where $C_{12} \equiv \psi_1\psi_2^*$. By writing space and time coordinates as follows:

$$\begin{aligned}\mathbf{r}_1 &= \mathbf{r} - \mathbf{s}/2, \quad t_1 = t - \tau/2\\ \mathbf{r}_2 &= \mathbf{r} + \mathbf{s}/2, \quad t_2 = t + \tau/2\end{aligned} \qquad [7.11]$$

Equation [7.10] can be rewritten as follows:

$$\left[\frac{\hbar^2}{m}\frac{\partial}{\partial \mathbf{r}}\cdot\frac{\partial}{\partial \mathbf{s}} - i\hbar\frac{\partial}{\partial t}\right]C_{12} = -(V_1 - V_2)C_{12} \qquad [7.12]$$

By developing potential $V_j$ in a Taylor series around $V(\mathbf{r},t)$, we can easily show (see Chapter 3):

$$(V_1 - V_2) = 2\sinh\left(\frac{\mathbf{s}}{2}\cdot\frac{\partial}{\partial \mathbf{s}} + \frac{\tau}{2}\frac{\partial}{\partial t}\right)V(\mathbf{s},t) \qquad [7.13]$$

By introducing a double Fourier transform of quantity $C_{12} \equiv C(\mathbf{r},\mathbf{s},t,\tau)$ for variables $\mathbf{s}$ and $\tau$, we have:

$$W(\mathbf{r},t,\omega,\mathbf{k}) = \int d\mathbf{s}\int d\tau C(\mathbf{r},\mathbf{s},t,\tau)\exp(-i\mathbf{k}\cdot\mathbf{s} + i\omega\tau) \qquad [7.14]$$

or again, in terms of the wave function $\psi$, we have:

$$W(\mathbf{r},t,\omega,\mathbf{k}) = \int d\mathbf{s} \int d\tau \psi(\mathbf{r}+\mathbf{s}/2,t+\tau/2)\psi^*(\mathbf{r}-\mathbf{s}/2,t-\tau/2)\exp(-i\mathbf{k}\cdot\mathbf{s}+i\omega\tau) \qquad [7.15]$$

which is none other than the Wigner distribution associated with the Bose–Einstein condensate. Using the Wigner distribution [7.15], the evolution equation [7.12] can be rewritten as:

$$\left(\frac{\hbar^2}{m}\mathbf{k}\cdot\frac{\partial}{\partial\mathbf{r}} - \hbar\frac{\partial}{\partial t}\right) W = -2V\sin\Lambda' W \qquad [7.16]$$

where the operator $\Lambda'$ is given by:

$$\Lambda' = \frac{1}{2}^{\leftarrow}\left(\frac{\partial}{\partial\mathbf{r}}\cdot\frac{\partial}{\partial\mathbf{k}} - \frac{\partial}{\partial t}\frac{\partial}{\partial\omega}\right)^{\rightarrow} \qquad [7.17]$$

In this definition, arrows show in what direction derivatives act, that is on $V$ or on $W$.

Equation [7.16] can be simplified by observing that the Gross–Pitaevskii equation implies a relationship between energy and momentum. Thus, $\omega$ must be a function of $\mathbf{k}$ ($\omega = \omega(\mathbf{k})$). We can then express this as:

$$W(\mathbf{r},t,\omega,\mathbf{k}) = W(\mathbf{r},\mathbf{k},t)\delta(\omega - \omega(\mathbf{k})) \qquad [7.18]$$

This leads to a simpler equation for $W(\mathbf{r},\mathbf{k},t)$. In fact, equation [7.16] is reduced to:

$$\left(\frac{\hbar^2}{2m}\mathbf{k}\cdot\nabla - i\hbar\frac{\partial}{\partial t}\right) W = -2\left(V_0 + g|\psi|^2\right)(\sin\Lambda)W \qquad [7.19]$$

where:

$$\Lambda = \frac{1}{2}\frac{\overleftarrow{\partial}}{\partial\mathbf{k}}\cdot\frac{\overrightarrow{\partial}}{\partial\mathbf{r}} \qquad [7.20]$$

It can be observed that the only difference between equation [7.19], established from the Gross–Pitaevskii equation, and equation [3.58], established in Chapter 3 from the linear Schrödinger equation, is the presence of the term $g|\psi|^2$. This does not depend on the Wigner distribution $W$, but on the wave function of condensate $\psi$. Kinetic equation [7.19] can be used in the same way as the Wigner–Moyal kinetic equation [3.58] using the usual perturbative processing. We will illustrate this in section 7.3 with the study of Bogoliubov oscillations.

## 7.3. Kinetic approach to Bogoliubov oscillations

The kinetic approach can be used to study the speed of sound in a Bose–Einstein condensate (Mendonça et al. 2005). For this, consider kinetic equation [7.19]. It can be written as follows:

$$\left(\frac{\hbar^2}{2m}\mathbf{k}\cdot\nabla - i\hbar\frac{\partial}{\partial t}\right)W = -2V\sin\Lambda W \qquad [7.21]$$

where the operator $\Lambda$ acts on the right on the Wigner distribution $W$ and on the left on potential $V$ (see equation [7.20]). The potential can be written as:

$$V = V_0 + g\int W(\mathbf{r},\mathbf{k},t)\frac{d\mathbf{k}}{(2\pi)^3} + \delta V \qquad [7.22]$$

with:

$$\delta V = g\left(|\psi(\mathbf{r},t)|^2 - \int W(\mathbf{r},\mathbf{k},t)\frac{d\mathbf{k}}{(2\pi)^3}\right) \qquad [7.23]$$

Equation [7.21] is the kinetic representation of the Gross–Pitaevskii equation. However, it is of little use as it stands. We will have to make some simplifying assumptions to be able to draw a physical description of the condensate. It may be interesting to consider the case of a potential that varies slowly. In this case, it is possible to ignore derivatives of potential of order greater than 1 and write $sin\Lambda \approx \Lambda$ (quasi-classic approximation). The fluctuating part of the potential, in this case, can also be ignored, $\delta V = 0$.

With these simplifying hypotheses, equation [7.21] is simply reduced to:

$$\left(\frac{\partial}{\partial t} + \mathbf{v}\cdot\nabla + \mathbf{F}\cdot\frac{\partial}{\partial \mathbf{k}}\right)W = 0 \qquad [7.24]$$

where $\mathbf{v} = \hbar\mathbf{k}/m$ is the velocity of atoms of the condensate corresponding to wave vector $\mathbf{k}$ and $\mathbf{F} = -\nabla V$ is the force derived from potential $V$. Thus, the nonlinear term of the Gross–Pitaevskii equation is hidden behind force $\mathbf{F}$. In this form, equation [7.24] is none other than an equation for conservation of the $W$ distribution in phase space. It can be written in the form of Liouville's equation:

$$\frac{d}{dt}W(\mathbf{r},\mathbf{k},t) = 0 \qquad [7.25]$$

To apply these results, it may be advantageous to consider the one-dimensional case of a beam moving on the $z$ axis and to neglect the axial variation of potential $V_0$ ($\partial V_0/\partial z \sim 0$) (Mendonça et al. 2005). Kinetic equation [7.24] can then be written as:

$$\left(\frac{\partial}{\partial t} + v_z\frac{\partial}{\partial z} + F_z\frac{\partial}{\partial k}\right) W(z,k,t) = 0 \qquad [7.26]$$

where velocity $v_z$ and force $F_z$ are expressed as:

$$v_z = \frac{\hbar k}{m} + g\frac{\partial}{\partial t}I(z,t)$$
$$F_z = \frac{dk}{dt} = -g\frac{\partial}{\partial z}I(z,t) \qquad [7.27]$$

with $I(z,t)$ representing the intensity of the condensate beam, defined by:

$$I(z,t) = \int W(z,k,t)\frac{dk}{2\pi} \qquad [7.28]$$

By linearizing kinetic equation [7.26] around an equilibrium distribution $W_0(z,k,t)$, we have:

$$\left(\frac{\partial}{\partial t} + v_z\frac{\partial}{\partial z}\right) \tilde{W}(z,k,t) + \tilde{F}\frac{\partial}{\partial k}W_0(z,k,t) = 0 \qquad [7.29]$$

where we get the force perturbation from:

$$\tilde{F} = -\frac{g}{\hbar}\frac{\partial}{\partial z}\tilde{I}(z,t) = -\frac{g}{\hbar}\frac{\partial}{\partial z}\int \tilde{W}(z,k,t)\frac{dk}{2\pi} \qquad [7.30]$$

Considering perturbations in the form of plane waves, $\tilde{W},\tilde{I} \sim \exp(ikz - i\omega t)$, equation [7.29] gives a link between the amplitude of the perturbation of distribution $\tilde{W}$ and the perturbation of intensity $\tilde{I}$:

$$\tilde{W} = -\frac{gk}{\hbar(\omega - kv')}\tilde{I}\frac{\partial}{\partial k'}W_0(k') \qquad [7.31]$$

where the wavenumber of particles is denoted as $k'$ to avoid confusion with the wavenumber of the perturbation $k$. By integrating the preceding relationship on $k'$, we have:

$$1 + \frac{g}{\hbar}k\int \frac{\partial W_0(k')/\partial k'}{(\omega - \hbar kk'/m)}\frac{dk'}{2\pi} = 0 \qquad [7.32]$$

This is the dispersion relation of an axial perturbation in a condensate.

To have a physical description, let us first consider the case of a zero temperature or, in other words, of a zero velocity dispersion of equilibrium distribution $W_0$. Within this limit, there are no kinetic effects per se and we should find results established in the context of the hydrodynamic representation (see section 7.1). In this case, the distribution is simply reduced to a Dirac delta, namely:

$$W_0\left(k'\right) = 2\pi n_0 \delta\left(k' - k'_0\right) \qquad [7.33]$$

where:

$$n_0 = \int W_0\left(k'\right) dk'/2\pi \qquad [7.34]$$

is the density of atoms in the condensate. With distribution [7.33], integral dispersion relation [7.32] is simply reduced to:

$$1 - \frac{gk^2}{m}\frac{n_0}{\left(\omega - kv'_0\right)^2} = 0 \qquad [7.35]$$

where $v'_0 = \hbar k'_0/m$ is the velocity of the beam. Relationship [7.35] can also be written as:

$$\left(\omega - kv'_0\right)^2 = k^2 c_s^2 \qquad [7.36]$$

where:

$$c_s = \sqrt{gn_0/m} \qquad [7.37]$$

is the Bogoliubov speed of sound in the condensate. This is none other than the dispersion relation obtained in the hydrodynamic context, that is, equation [7.7] (for atoms initially at rest, $v'_0 = 0$). Indeed, by associating the Wigner distribution with a Dirac delta, we implicitly make the hypothesis of a zero temperature or absence of kinetic effects.

Purely kinetic effects, such as the Landau damping discussed in section 5.3 in the context of plasmas, can however be processed with the help of the kinetic equation [7.26]. To see this, consider a situation where the distribution does not correspond to a Dirac delta but has a certain width (a moment of order 2, $\langle v^2 \rangle$ that is not zero). Thus, we consider the case where a non-zero number of particles has a velocity close to the phase velocity $c_s$ (resonant particles). Energy exchanges between these particles and the oscillation give rise to the Landau damping phenomenon. In this case, by retaining this contribution in the integral dispersion relation [7.32] (see equation [5.40] in the case of plasmas), we have:

$$1-\frac{k^2c_s^2}{\omega^2}-\frac{i}{2}\frac{gm}{\hbar^2}\left(\frac{\partial W_0}{\partial k'}\right)_{k'=k'_s}=0 \qquad [7.38]$$

where $k'_s = mc_s/\hbar$. The imaginary part in relation [7.38] is responsible for damping oscillations. By denoting $\omega_i$ as the imaginary part of the frequency (i.e. $\omega = c_s k + i\omega_i$), and considering the case of low damping ($|\omega_i| \ll c_s k$), we have:

$$\omega_i = \frac{\omega}{4}\frac{gm}{\hbar^2}\left(\frac{\partial W_0}{\partial k'}\right)_{k'=k'_s} \qquad [7.39]$$

This expression represents Landau damping (collisionless damping) of Bogoliubov oscillations in a Bose–Einstein condensate.

# 8

# Cosmology and Dark Matter Models

*Everything speaks in the universe;*
*there is nothing that doesn't have its language.*

Jean de La Fontaine

In this chapter, we will discuss the application of kinetic and hydrodynamic methods in the context of cosmology and dark matter models. This chapter can be seen as a continuation of Chapter 6 on gravitational systems (gravitation being the dominant interaction on a cosmological scale). However, we will see here aspects that have been neglected in Chapter 6, such as the expansion of the Universe. Specifically, we will see how to include the expansion of the Universe coherently in hydrodynamic equations and how this affects the mechanism of Jeans gravitational instability. We will also study fluid and kinetic representations of a particular model of dark matter, namely *scalar field dark matter*.

## 8.1. Hydrodynamics of the Universe

Cosmology is a field where hydrodynamic equations are particularly well justified, even if they must be used with caution. Indeed, in cosmology, we are interested in the description of the Universe (observable) at very large scales of distance. At such scales, there is no need to enter into a microscopic, or even mesoscopic, description and the Universe as a whole can be modeled in the form of a fluid. Today, fluid models are ubiquitous in cosmology, both in theoretical studies and numerical simulations.

Historically, the usefulness of the fluid model in cosmology has not always been obvious. It is essentially due to Zel'dovich (1977) who was the first to provide a clear formulation of what he called the *hydrodynamics of the Universe*. In what follows, we will also essentially present its formulation. What will be done in this section is quite similar to what has been discussed in Chapter 6 devoted to the study of gravitational systems. Indeed, gravity is by far the dominant force on the cosmological (and even astrophysical) scale. The main new ingredient discussed in this section is the expansion of the Universe, which is essentially due to Friedmann. The expansion of the Universe refers to the phenomenon that sees, on a large scale, objects composing the Universe move away from each other. This spacing is interpreted as a swelling of space itself that forces astrophysical objects to move away from each other, rather than a displacement of these objects per se. It is to this phenomenon that we referred above, specifying that the hydrodynamic model must be used "with caution". Let us begin by writing hydrodynamic equations governing the evolution of a gravitational fluid, with an equation of state defining pressure $p(\rho,s)$ ($\rho$ being the mass density and $s$ the entropy), as follows:

$$\begin{aligned}&\frac{\partial \rho}{\partial t}+\nabla(\rho\mathbf{u})=0\\&\frac{\partial \mathbf{u}}{\partial t}+(\mathbf{u}\nabla)\mathbf{u}=-\nabla\varphi-\frac{1}{\rho}\nabla p\\&\frac{\partial s}{\partial t}+(\mathbf{u}\nabla)s=0\\&\nabla^2\varphi=4\pi G\rho\end{aligned} \quad [8.1]$$

In cosmology, we look for solutions in the following form:

$$\begin{aligned}&\rho=\rho(t),\quad p=p(t),\quad s=\text{ const}\\&\mathbf{u}=H\mathbf{r},\quad \varphi=\frac{2\pi G}{3}\rho(t)r^2\end{aligned} \quad [8.2]$$

where $H$ is the *Hubble parameter*. This is the constant of proportionality between the distance and the apparent velocity of galaxies in the observable Universe (the farther away a galaxy is from Earth[1], the faster it moves away from it). The solutions sought [8.2] originate from the fact that, once replaced in the hydrodynamic model [8.1], they involve the following laws, determining the evolution of $\rho(t)$ and $H(t)$:

$$\frac{d\rho}{dt}=-3H\rho,\quad \frac{dH}{dt}+H^2=-\frac{4\pi G\rho}{3} \quad [8.3]$$

These are *Friedmann–Lemaître equations*. These correspond to equations of general relativity written in the case of a *homogeneous* and *isotropic* cosmological model (represented by a Robertson–Walker metric).

1 This is true not only for Earth, but for any point in space from which an observation would be made.

Let us analyze conditions [8.2] to show that they describe a homogeneous and isotropic universe. First, these equations indicate that density $\rho(t)$ and entropy $s$ do not depend on the space coordinate $\mathbf{r}$: this reflects homogeneity. As for velocity, the condition of homogeneity is a little subtle because of Hubble's law, which precisely indicates that velocity is not the same everywhere. However, we must remember that velocity is always relative. Thus, it is always possible to add a vector to velocity without anything changing. We can thus make a transverse displacement of the space coordinate $\mathbf{r}$, provided that we also consider the transformation of velocity, due to the change of reference frame:

$$\mathbf{r}' = \mathbf{r} - \mathbf{r}_0, \quad \mathbf{u}' = \mathbf{u} - H\mathbf{r}_0 \tag{8.4}$$

By replacing these relationships in equation [8.2], we have:

$$\mathbf{u}' = H\mathbf{r}' \tag{8.5}$$

Thus, an observer located at $\mathbf{r_0}$ will see the same "movement", due to the expansion of the Universe, as an observer located at $\mathbf{r} = 0$. In the same way, it is possible to understand the homogeneity of potential $\varphi$. According to conditions [8.2], we have $\varphi \sim \mathbf{r}^2$, which implies the following initial conditions:

$$r = 0, \quad \varphi = 0$$

and: [8.6]

$$r = 0, \quad \nabla\varphi = 0$$

If a transverse displacement of the space coordinate ($\mathbf{r}' = \mathbf{r} - \mathbf{r_0}$) did not affect the potential, we would have:

$$\varphi = \frac{2\pi}{3}G\rho r'^2 + \frac{4\pi G}{3}\rho\left(\mathbf{r}'\mathbf{r}_0\right) + \frac{2\pi G}{3}\rho r_0^2 \tag{8.7}$$

this seems to violate the homogeneity condition. However, the potential is defined to within one constant and the last term of equation [8.7] can be discarded. This is true for other potentials, but for the gravitational potential, it is also necessary to take into account the fact that a gravitational field is equivalent to an acceleration (this is *Einstein's equivalence principle*, see Figure 8.1). As a result, by changing the frame of reference and doing the appropriate transformations, we have:

$$\varphi' = \frac{2\pi G}{3}\rho r'^2 \tag{8.8}$$

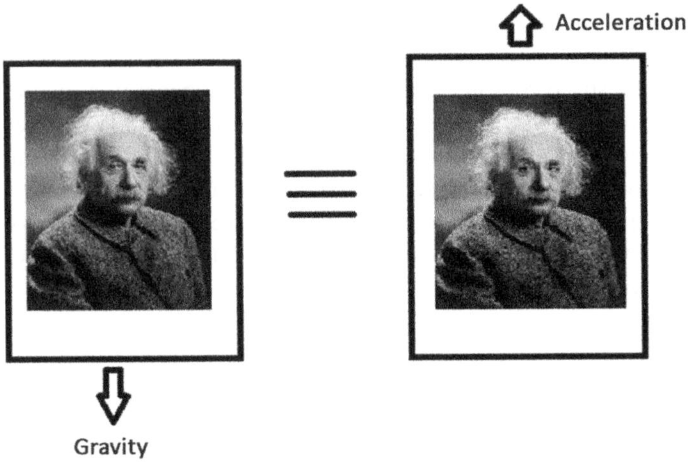


**Figure 8.1.** *Illustration of the equivalence principle: the situation of a frame of reference moving at an acceleration* **g** *is equivalent to the presence of a gravitational potential* **gr**

This shows that homogeneity is well preserved. Note that isotropy is also assumed by writing the Poisson equation. As noted by Zel'dovich (1977), homogeneity and isotropy conditions not only represent a useful hypothesis that would facilitate processing, but also result from a large number of astrophysical observations.

Let us now return to the system of equations [8.3], which can be solved trivially. In fact, by eliminating time $t$ from the equations, we have:

$$\frac{dH^2}{d\rho} = \frac{2}{3}\frac{H^2}{\rho} + \frac{8\pi G}{9}, \quad H^2 = \frac{8\pi G\rho}{3} + b\rho^{2/3} \qquad [8.9]$$

where $b$ is an integration constant. Using the current value of the Hubble constant, constant $b$ can be determined as follows:

$$H_0^2 = \frac{8\pi G}{3}\rho_0 + b\rho_0^{2/3}$$

$$b = \frac{H_0^2}{\rho_0^{2/3}}(1-\Omega), \quad \Omega = \rho/\rho_c \qquad [8.10]$$

$$\rho_c = 3H_0^2/8\pi G$$

where $\Omega$ is the ratio between the current density $\rho_0$ and a critical density $\rho_c$, constructed from the current Hubble constant $H_0$. Note that from the previous equations, we have:

$$\Omega > 1 \Rightarrow b < 0 \qquad [8.11]$$

In this case, the Hubble parameter $H$ does not change sign and expansion continues indefinitely. However:

$$\Omega < 1 \Rightarrow b > 0 \qquad [8.12]$$

In this case, for a density $\rho = (3|b|/8\pi G)^3$, Hubble parameter $H$ is canceled out. As a result, expansion stops and becomes a compression.

For $\Omega = 1$ and $b = 0$, the equations are particularly simple, and we have:

$$\rho = \frac{1}{6\pi G t^2}, \quad H = \frac{2}{3t}, \quad a \propto t^{2/3} \qquad [8.13]$$

where we defined the *linear scale*, that is, $a = \exp \int H dt$. The distance between any two points in space is proportional to $a$.

As noted by Zel'dovich (1977), the previous result could have been obtained in a more elementary way, simply by considering a finite sphere of radius $a$ and constant density, taking into account Hubble's law $\mathbf{u} = H\mathbf{r} = (\mathbf{r}/a)(da/dt)$. Equations of motion are then reduced to:

$$\frac{d^2a}{dt^2} = -\frac{GM}{a^2}, \quad M = \frac{4\pi}{3}\rho a^3 = \text{const} \qquad [8.14]$$

### 8.1.1. *Perturbation in a static Universe*

The focal point of this section is the inclusion of the expansion of the Universe. Thus, in comparison with what has been presented in Chapter 6, equilibrium quantities can no longer be considered constant. However, before describing the evolution of perturbations in an expanding universe, it may be useful to briefly return to the static case and clarify some aspects.

Let us then take the system of fluid equations [8.1] and introduce perturbations around equilibrium values as follows:

$$\begin{aligned}
&\rho = \rho_0 \left[1 + \delta_k \exp(i\mathbf{kr})\right] \\
&\mathbf{u} = \mathbf{u}_k \exp(i\mathbf{kr}) \\
&s = s_0 \\
&\varphi = \varphi_k \exp(i\mathbf{kr}) \\
&p = p_0 + c^2 \rho_0 \delta_k \,\mathrm{cxp}(i\mathbf{kr}) \\
&c^2 = (\partial p/\partial \rho)_{s=s_0}
\end{aligned} \qquad [8.15]$$

Quantities containing index $k$ represent perturbations, which are assumed to be small compared with equilibrium quantities. After linearization, we have:

$$\begin{aligned}
&\dot{\delta}_k = -i\left(\mathbf{k}\mathbf{u}_k\right), \quad k^2 \varphi_k = -4\pi G \rho_0 \delta_k \\
&\dot{\mathbf{u}}_k = -i\mathbf{k}\left(\varphi_k + c^2 \delta_k\right)
\end{aligned} \qquad [8.16]$$

where points on variables represent derivatives with respect to $t$. Looking only at longitudinal perturbations ($\mathbf{u_k} = v_k \mathbf{k}/|\mathbf{k}|$), we have:

$$\begin{aligned}
&\dot{\delta}_k = -i v_k k \\
&\dot{v}_k = -ik\left(-4\pi G \rho_0/k^2 + c^2\right)\delta_k
\end{aligned} \qquad [8.17]$$

These equations have an exponential solution, that is:

$$\delta_k = \delta_{0k} \exp(\gamma t), \quad v_k = v_{0k} \exp(\gamma t) \qquad [8.18]$$

where:

$$\gamma = \pm\sqrt{4\pi G \rho_0 - c^2 k^2} \qquad [8.19]$$

This result was obtained in Chapter 6 dealing with the study of Jeans instability. Solutions [8.18] with [8.19] show the existence of two modes: one decreases exponentially with time and the other increases exponentially with time. The latter represents the Jeans gravitational instability.

One point we had not stressed enough, but which is becoming crucial in cosmology, is the *exponential* evolution of perturbations, which has turned out to no longer be true in the case of an expanding universe. The incorporation of the expansion of the Universe into the previously described fluid equations will be discussed in section 8.1.2.

It should also be noted that in the linear approach, there is a principle of superposition. Any linear combination of solutions is also a solution. Or:

$$\rho = \rho_0 \left[1 + \sum \delta_k \exp\left(\gamma_k t + i\mathbf{kr}\right)\right] \quad [8.20]$$

Thus, the two modes coexist.

### 8.1.2. *Perturbations in an expanding Universe*

The missing ingredient in the previous discussion is the expansion of the Universe, as up until then perturbations around stationary equilibrium values were considered. For density, for example, perturbations were introduced around an equilibrium value $\rho_0$ assumed to not depend on time.

This approach is no longer valid if we take into account the expansion of the Universe which implies that equilibrium density changes over time, even independently of any perturbation of matter, due to the stretching of space itself. The same is true for the equilibrium velocity which can no longer be considered stationary because of Hubble's law.

To deal with matter perturbations in an expanding universe, let us take the system of fluid equations [8.1] and introduce perturbations around non-stationary values as follows:

$$\begin{aligned} \rho &= \rho_0(t)\left[1 + \delta_k(t)\exp(i\mathbf{kr})\right] \\ \mathbf{u} &= H(t)\mathbf{r} + \mathbf{u}_k \exp(i\mathbf{kr}) \\ \varphi &= \frac{2\pi G}{3}\rho_0 r^2 + \varphi_k \exp(i\mathbf{kr}) \end{aligned} \quad [8.21]$$

where $\rho_0(t)$ and $H(t)$ satisfy Friedmann–Lemaître equations [8.3], describing a homogeneous and isotropic universe. Solutions in the form of [8.21] are only possible in the case of a wave vector **k**, itself varying in time, that is:

$$\frac{d\mathbf{k}}{dt} = -H\mathbf{k} \quad [8.22]$$

This means that wavenumber $k \equiv |\mathbf{k}|$ decreases over time or, in other words, the wavelength of perturbations increases over time, due to the expansion of the Universe, proportionally to scale factor $a(t)$.

Introducing the so-called Lagrangian coordinates $\sigma(\xi,\eta,\zeta)$ for unperturbed solutions, we have:

$$\mathbf{r} = a(t)\sigma \quad [8.23]$$

The exponent of perturbations is then written as:

$$\exp(i\mathbf{kr}) = \exp(i\kappa\sigma), \quad \kappa = a(t)\mathbf{k} \quad [8.24]$$

Thus, solutions have a constant so-called Lagrangian wave vector $\kappa$, but a so-called Eulerian wave vector $\mathbf{k}$, which depends on time through scale factor $a(t)$. With this subtlety, what was discussed previously in the stationary case remains valid. Perturbations with a wavenumber greater than the Jeans number behave like simple oscillations, while perturbations with a wavenumber lower than the Jeans number give rise to instability. However, what changes is the evolution of perturbations over time, which is no longer exponential as in the stationary case (see equation [8.18]).

For a "flat" universe (equation [8.13]), we have the two modes of perturbation:

$$\begin{aligned} &\delta_+ \sim t^{2/3}, u_+ \sim t^{1/3} \\ &\delta_- \sim t^{-1}, u_- \sim t^{-4/3} \end{aligned} \quad [8.25]$$

Here again, the linearity of equations makes it possible to write in a general way:

$$\begin{aligned} &\delta = t^{2/3}\delta_+(\mathbf{r}/a) + t^{-1}\delta_-(\mathbf{r}/a) \\ &\mathbf{u} = t^{1/3}\mathbf{u}_+(\mathbf{r}/a) + t^{-4/3}\mathbf{u}_-(\mathbf{r}/a) + t^{-2/3}\mathbf{u}_{\text{vort}}(\mathbf{r}/a) \end{aligned} \quad [8.26]$$

taking into consideration, however, the change in the wave vector and, where:

$$\nabla \times \mathbf{u}_+ = \nabla \times \mathbf{u}_- = 0, \quad \nabla \cdot \mathbf{u}_+ = -\frac{2}{3}\frac{\delta_+}{t}, \quad \nabla \cdot \mathbf{u}_- = \frac{\delta_-}{t}, \quad \nabla \cdot \mathbf{u}_{\text{vort}} = 0 \quad [8.27]$$

It should be noted that if all the modes have the same order of magnitude at an instant $t_0$, only the mode increasing in time survives for $t \gg t_0$.

Likewise, because of the linearity, we can write:

$$\begin{aligned}
\rho &= \frac{1}{6\pi G t^2}\left[1 + t^{2/3}\delta_+(\mathbf{r}/a)\right] \\
\mathbf{u} &= \frac{2\mathbf{r}}{3}\frac{\mathbf{r}}{t} + t^{1/3}\mathbf{u}_+(\mathbf{r}/a) \\
a &\propto t^{2/3}
\end{aligned} \qquad [8.28]$$

Thus, the important point, revealed by the introduction of the expansion of the Universe, is that perturbations do not evolve exponentially $\sim \exp(\gamma t)$, but follow a power law $\sim t^{2/3}$.

As observed by Zel'dovich (1977), such a change in $t$-power can be qualitatively inferred from standard Jeans processing (see Chapter 6). For this, it is enough to note that in the standard case, we have $\gamma = \pm\sqrt{4\pi G \rho_0}$. However, in an expanding universe, the equilibrium density varies according to $\rho_0 = 1/6\pi G t^2$.

It is then possible to define $\tilde{\gamma}(t)$ as an average of $\gamma$, that is:

$$\tilde{\gamma}(t) = \frac{1}{t}\int \gamma dt = \sqrt{\frac{2}{3}}\frac{\ln t}{t} \qquad [8.29]$$

Perturbations then become proportional to:

$$\exp(\pm\tilde{\gamma}t) = \exp\left(\pm\sqrt{\frac{2}{3}}\ln t\right) = t^{\pm\sqrt{2/3}} \qquad [8.30]$$

There is indeed a potential law although, quantitatively, this gives an incorrect exponent ($\sqrt{2/3} \approx 0.85$ rather than $2/3 \approx 0.67$).

## 8.2. Dark matter models

The final area of application will be dark matter models. In particular, we will focus on a particular model where dark matter is seen as a gas of ultra-light bosons interacting via gravity forces (Magaña and Matos 2012). Before presenting the model, it is important to mention the concept of dark matter and the indications in favor of its existence. The main observational indicator in favor of the existence of dark matter comes from the fact that the observed behavior of galaxies does not seem to be in line with theoretical predictions. For galaxies to behave as the observations indicate, they would have to contain much more matter. Since this material is invisible to telescopes, it is called *dark matter*.

At present, there is no direct empirical evidence for the existence of dark matter, although indirect observations do indeed favor its existence. Several dark matter models have been proposed and theoretical predictions of these models are constantly compared with observations. The standard model of cosmology suggests that dark matter would be cold (*cold dark matter*): this means that the effects of pressure forces are weak. This model is in line with observations on a very large scale (cosmological scale). At smaller scales, however, as at the galaxy scale, the model does not appear to be supported by observations (Chavanis 2020).

One possible solution to this problem is to introduce a mechanism (similar to pressure) that could counterbalance the effect of gravity at small scales. We just saw such an effect in section 6.1. We have indeed seen that Bohm's quantum potential acts against the attractive forces of gravity, making matter more stable. Thus, a dark matter model, which would correct the predictions of cold dark matter, would be a model where quantum effects play a role. This model suggests that dark matter is formed from a Bose–Einstein condensate, where bosons (thought to be ultra-light) interact through gravitational interactions.

Such a model crystallizes, in a sense, what we saw in Chapters 6 and 7, namely the study of gravitational systems and Bose–Einstein condensates. The starting equations will, therefore, be the Schrödinger equation, where a nonlinearity term occurs due to interactions between bosons, that is, the Gross–Pitaievskii equation [7.1], coupled with the Poisson equation for gravitational potential (this actually represents the Newtonian limit, where the effects of general relativity are not taken into account). Thus, we have:

$$
\begin{aligned}
&i\hbar\frac{\partial\Psi(\mathbf{r},t)}{\partial t} = \left[-\frac{\hbar^2}{2m}\nabla^2 + U(\mathbf{r}) + g|\Psi(\mathbf{r},t)|^2 + m\Phi\right]\Psi(\mathbf{r},t) \\
&\nabla^2\Phi = 4\pi G|\Psi(\mathbf{r},t)|^2
\end{aligned}
\qquad [8.31]
$$

The system of equations can be used in its hydrodynamic or kinetic formulation (see Chapter 3).

### 8.2.1. *Hydrodynamic representation*

Introducing the Madelung transformation[2]:

$$\Psi = \sqrt{\rho} e^{iS/\hbar} \quad \mathbf{v} = \frac{\nabla S}{m} \tag{8.32}$$

system [8.31] can be rewritten in hydrodynamic form as follows:

$$\begin{aligned} &\frac{\partial \rho}{\partial t} + \nabla \cdot (\rho \mathbf{v}) = 0 \\ &\frac{\partial \mathbf{v}}{\partial t} + (\mathbf{v} \cdot \nabla)\mathbf{v} + \nabla \Phi + \frac{\nabla p}{\rho} - \frac{\hbar^2}{2m^2} \nabla \left( \frac{\nabla^2 \sqrt{\rho}}{\sqrt{\rho}} \right) = 0 \end{aligned} \tag{8.33}$$

where we defined:

$$p = \frac{1}{2} g \rho^2 \tag{8.34}$$

Considering small perturbations around stationary equilibrium values:

$$\rho = \rho_0, \quad \mathbf{v} = 0, \quad \Phi = 0 \tag{8.35}$$

and by seeking a solution in the form of a plane wave, the following dispersion relation is obtained:

$$\omega^2 = \left( \frac{\hbar^2 k^2}{4m^2} + c_s^2 \right) k^2 - 4\pi G \rho_0 \tag{8.36}$$

where $c_s^2 \equiv g$. Dispersion relation [8.36] indicates that there are two effects that oppose gravitational instability, namely the term $c_s^2$ describing the interaction in bosons, and the quantum effects due to the Bohm quantum potential. Relationship [8.36] is in fact formally identical to relationship [6.11], where $c_s^2$ plays the role of thermal pressure $3k_B T/m$.

In the case where the interaction between bosons is weak, term $c_s^2$ can be ignored. In this case, we can find the value of $k$, distinguishing between real $\omega$ and imaginary $\omega$, from dispersion relation [8.36]. This is given by:

$$k_J = \frac{1}{2\pi} \left( \frac{m^2 G \rho_0}{\pi^3 \hbar^2} \right)^{1/4} \tag{8.37}$$

2 To simplify, the wave function is normalized to *mass* density $\rho$, that is, $|\Psi(\mathbf{r},t)|^2 \equiv \rho = \mathrm{mn}$.

This is the Jeans number for a dark matter pattern in the form of Bose–Einstein condensate without interaction.

From [8.37], we can also define the Jeans length:

$$\lambda_J = \frac{2\pi}{k} = \left(\frac{\pi^3\hbar^2}{m^2 G\rho_0}\right)^{1/4} \qquad [8.38]$$

as well as the Jeans mass, namely the mass initially contained in a sphere of radius $\lambda_J$, that is:

$$M_J = \frac{4\pi}{3}\rho_0\lambda_J^3 = \frac{4}{3}\pi^{\frac{12}{4}} G^{-\frac{1}{4}}\hbar^{\frac{2}{2}} m^{-\frac{2}{2}}\rho_0{}^{\frac{1}{4}} \qquad [8.39]$$

This represents the maximum mass that a dark matter halo can have. Beyond this mass, dark matter, like visible (baryonic) matter, tends to fragment, forming smaller halos.

As discussed in section 8.1, an important ingredient of cosmology is not considered in the previous discussion, namely the expansion of the Universe. Thus, the equilibrium values we have considered, and in particular the value of density at equilibrium $\rho_0$ is not constant, but decreases due to the expansion of the Universe. However, the expressions obtained previously remain valid, provided that $\rho_0$ is no longer considered as static, but as a function of the scale factor. In cosmology, we write:

$$\begin{aligned} k_J(z) &= \frac{1}{2\pi}\left(\frac{m^2 G\rho_0(z)}{\pi^3\hbar^2}\right)^{1/4} \\ \lambda_J(z) &= \left(\frac{\pi^3\hbar^2}{m^2 G\rho_0(z)}\right)^{1/4} \\ M_J(z) &= \frac{4}{3}\pi^{\frac{12}{4}} G^{-\frac{1}{4}}\hbar^{\frac{2}{2}} m^{-\frac{2}{2}}\rho_0(z)^{\frac{1}{4}} \end{aligned} \qquad [8.40]$$

where $z$ is the spectral shift, related to scale factor $a(t)$ by:

$$1+z = \frac{a_0}{a(t)} \qquad [8.41]$$

### 8.2.2. *Kinetic representation*

Note that it is also possible to describe the gravitational instability of dark matter through kinetic theory using the Wigner–Moyal theory. Also, we will present a contribution of the author (Ourabah 2020b) in this context here. To simplify calculations, we will ignore interactions between bosons ($g = 0$). The Wigner–Moyal equation is to be considered here:

$$\left(\frac{\partial}{\partial t} + \frac{\mathbf{p}}{m}\frac{\partial}{\partial \mathbf{r}}\right) W + \frac{2m}{\hbar}\Phi \sin\left(\frac{\hbar}{2}\frac{\overleftarrow{\partial}}{\partial \mathbf{r}}\frac{\overrightarrow{\partial}}{\partial \mathbf{p}}\right) W = 0 \qquad [8.42]$$

with $\mathbf{p} = \hbar\mathbf{k}$ and we introduced gravitational potential $\Phi$. The Wigner–Moyal equation will be coupled to the Poisson equation:

$$\nabla^2\Phi = 4\pi G \int d\mathbf{p}W \qquad [8.43]$$

where we used the fact that the integral of the Wigner distribution over the momentum (or velocities) space corresponds to density (here, mass density $\rho$). Considering perturbations of the Wigner distribution and potential in the form of plane waves:

$$\begin{aligned} W(x,p;t) &\equiv W_0(p) + \tilde{W}\exp[i(\mathbf{k}\cdot\mathbf{r} - \omega t)] \\ \phi(x,t) &\equiv \tilde{\phi}\exp[i(\mathbf{k}\cdot\mathbf{r} - \omega t)] \end{aligned} \qquad [8.44]$$

where $|\tilde{W}| \ll |W_0|$, we get:

$$\begin{aligned} &-i\left(\omega - \frac{p}{m}k\right)\tilde{W} + \frac{2m}{\hbar}\tilde{\phi}\sin\left(\frac{i\hbar k}{2}\frac{\vec{\partial}}{\partial p}\right) W_0 = 0 \\ &-k^2\tilde{\phi} = 4\pi G \int dp\tilde{W} \end{aligned} \qquad [8.45]$$

Combining equation [8.45] with the Poisson equation [8.43], we get:

$$\frac{8\pi i G m^3}{\hbar k^3}\int_{-\infty}^{+\infty} \frac{dp}{p - m\omega/k}\sin\left(\frac{i\hbar k}{2}\frac{\vec{\partial}}{\partial p}\right) W_0 - 1 = 0 \qquad [8.46]$$

This can be simplified by observing that the Wigner distribution has the following property:

$$2\sin\left(\frac{i\hbar k}{2}\frac{\vec{\partial}}{\partial p}\right)W_0(p) = i\left[W_0(p+\hbar k/2) - W_0(p-\hbar k/2)\right] \qquad [8.47]$$

Thus, we have:

$$1+\frac{4\pi G m^3}{\hbar k^3}\int_{-\infty}^{+\infty} dp \frac{W_0(p+\hbar k/2)-W_0(p-\hbar k/2)}{p-\omega m/k} = 0 \qquad [8.48]$$

This equation is the kinetic generalization of the dispersion relation obtained in the hydrodynamic context [8.36] (in the absence of interaction, $g = 0$). As the hydrodynamic model does not make any difference between the particles and assigns them a single velocity (fluid velocity), this can be found from the kinetic theory if the distribution is assimilated to a Dirac delta:

$$W_0(p) = \rho_0 \delta(p - p_0) \qquad [8.49]$$

In this case, relationship [8.48] gives the following dispersion relation:

$$\omega^2 = -4\pi G\rho_0 + \frac{\hbar^2 k^4}{4m^2} \qquad [8.50]$$

which is none other than the dispersion relation obtained through the hydrodynamic approach [8.36], within the considered limit $g = 0$, from which the Jeans wavenumber [8.37], as well as the associated wavelength and mass can be deduced.

However, the kinetic approach makes it possible to go further than the fluid approach. In particular, it makes it possible to include the effects of a non-zero temperature, by no longer associating the Wigner distribution at equilibrium, $W_0$, with a Dirac delta, but with a thermal distribution with a certain width (see, for example, the discussion in section 7.3), thus bridging the gap with other dark matter models where dark matter is assumed to be thermal (Hammann and Hasenkamp 2013; Lovell et al. 2014; Ourabah 2020a).

# Appendix

# Language of Relativity

In this appendix, we will present the elements of language related to special and general relativity. We will essentially deal with the notion of space-time and how to define objects in this space. These concepts are necessary for the formulation of kinetic and hydrodynamic approaches in the context of relativity, as presented in Chapter 4. More details can be found in Caroll (1997) and Rezolla and Zanotti (2013).

## A.1. Minkowski space-time

What we mean by *space-time* is simply a collection of events. An event $\mathscr{P}$ is thus represented by a point in space-time, defining a *place* and an *instant*, for a given observer.

## A.2. Coordinates

All coordinates that will be discussed here will be coordinates in an *inertial*, frame of reference, that is in a frame of reference that does not undergo acceleration or rotation.

It is possible to cover space-time using coordinates $\{x^{\mu}\}$ or $\left\{x^{\mu'}\right\}$, where $\mu, \mu' = 0,1,2,3$. The most trivial example of coordinate systems is given by *Cartesian* coordinates $(t,x,y,z)$, where $t$ represents time and $x,y,z$ represent space coordinates. In the same way, we can define *spherical* coordinates $(t,r,\theta,\phi)$. As the same system can be studied in one coordinate system or the other, it is possible to define a coordinate transformation $\{x^{\mu}\} \rightarrow \left\{x^{\mu'}\right\}$ making it possible to define the same points of space-time in another coordinate system.

Let us consider two events $\mathscr{P}_1$ and $\mathscr{P}_2$ in space-time, whose Cartesian coordinates are $(t_1, x_1, y_1, z_1)$ and $(t_2, x_2, y_2, z_2)$, respectively.

The *Minkowski space-time interval* is defined as follows[1]:

$$(\Delta s)^2 = -c^2(\Delta t)^2 + (\Delta x)^2 + (\Delta y)^2 + (\Delta z)^2 \qquad [A.2]$$

where $\Delta x^\mu := x_2^\mu - x_1^\mu$. The Minkowski space is a *pseudo-Euclidean* space. This is the space of the theory of special relativity.

Let us now consider a frame of reference $\mathscr{O}$ and a second frame of reference $\overline{\mathscr{O}}$ moving at a velocity $v$ relative to the first. Taking into consideration the invariance of the speed of light $c$, we have:

$$(\Delta s)^2 = (\Delta \bar{s})^2 \qquad [A.3]$$

It is said that the space-time interval is *invariant*. Since the speed of light $c$ is invariant, it will be possible to place $c = 1$ by convention.

## A.3. Four-vector

To define notions such as displacement, velocity and so on, it is necessary to define, in a formal sense, the notion of vector in space-time. We will use the usual notation of a *Lorentz transformation*, namely:

$$\Lambda_\beta^{\bar{\alpha}} := \frac{\partial x^{\bar{\alpha}}}{\partial x^\beta} \qquad [A.4]$$

which represents a matrix transforming the coordinates of a vector in one coordinate system to another.

A *four-vector* $\mathbf{V}$ is a collection of four numbers $\{V^\alpha\}$, which are transformed from a reference frame $\mathscr{O}$ to a reference frame $\overline{\mathscr{O}}$ according to the transformation:

$$V^{\bar{\alpha}} = \Lambda_\beta^{\bar{\alpha}} V^\beta \qquad [A.5]$$

1 We can just as easily choose another convention for the signs and write:

$$(\Delta s)^2 = +c^2(\Delta t)^2 - (\Delta x)^2 - (\Delta y)^2 - (\Delta z)^2 \qquad [A.1]$$

In the same way, a *uniform* $\tilde{p}$ is a set of four numbers $\{p_\alpha\}$, which are transformed from a reference frame $\mathcal{O}$ to a reference frame $\overline{\mathcal{O}}$ according to the transformation:

$$p_{\bar{\beta}} = \Lambda^{\alpha}_{\bar{\beta}} p_\alpha \qquad [A.6]$$

In what follows, we will use the *Einstein notation*; thus, a repeated index expresses an implicit sum. For instance:

$$\begin{aligned} V^\alpha p_\alpha &= U^0 p_0 + U^1 p_1 + U^2 p_2 + U^3 p_3 \\ V^i p_i &= V^1 p_1 + V^2 p_2 + V^3 p_3 \end{aligned} \qquad [A.7]$$

By convention, Greek letters represent sums on $\{0,1,2,3\}$, either on the temporal coordinate and the space coordinates, while Latin letters represent sums on $\{1,2,3\}$, or only on space coordinates.

## A.4. Tensors

Having defined the concept of four-vector (which we will simply call "vectors" in the following), it may be useful to define operations with these vectors. The simplest example is to take two vectors and produce a scalar. In this case, for two vectors **a** and **b**, a tensor **T** can be defined so as to have:

$$\mathbf{T}(\mathbf{a},\mathbf{b}) = c \qquad [A.8]$$

where $c$ it is a scalar. Obviously, this operation does not define a tensor but is a simple example to understand.

Note that components of a vector **a** in a coordinate system can be represented in two ways: in a *contravariant* form $a^\mu$ or *covariant* form $a_\mu$. In the same way as for a vector, a tensor can be represented by its coordinates. The tensor of the example of equation [A.8] can thus be represented by a 16-element matrix. From $a^\mu$ and $b^\mu$, a tensor can be written in a contravariant form $T^{\mu\nu}$, while from $a_\mu$ and $b_\mu$, a tensor can be expressed in a covariant form $T_{\mu\nu}$. It is also possible to construct a tensor $T^\mu_\nu$ from $a_\mu$ and $b^\nu$.

However, the notion of tensor is not limited to two indices: it is thus possible to define higher order tensors.

One property of defining tensors is to describe how they transform between a coordinate system $\{x^{\mu}\}$ and another coordinate system $\left\{x^{\mu'}\right\}$. Thus, we have covariant transformations:

$$\begin{aligned} T^{\bar{\mu}\bar{\nu}} &= \frac{\partial x^{\bar{\mu}}}{\partial x^{\mu}} \frac{\partial x^{\bar{\nu}}}{\partial x^{\nu}} T^{\mu\nu} \\ &= \Lambda^{\bar{\mu}}_{\mu} \Lambda^{\bar{\nu}}_{\nu} T^{\mu\nu} \end{aligned} \qquad [A.9]$$

## A.5. The metric

The scalar product between two four-vector **U** and **V** is given by:

$$\mathbf{U} \cdot \mathbf{V} := \mathbf{g}(\mathbf{U}, \mathbf{V}) = g_{\mu\nu} U^{\mu} V^{\nu} \qquad [A.10]$$

where **g** is the *metric*. The metric is the most important tensor in relativity. In the Minkowski space-time, components of the metric are usually denoted as $\eta_{\mu\nu}$. In the Cartesian coordinate system, we get elements of the Minkowski space-time metric from:

$$\eta_{\mu\nu} = \begin{pmatrix} -1 & 0 & 0 & 0 \\ 0 & 1 & 0 & 0 \\ 0 & 0 & 1 & 0 \\ 0 & 0 & 0 & 1 \end{pmatrix} \qquad [A.11]$$

In this case, matrices representing metrics $\eta_{\mu\nu}$ and $\eta^{\mu\nu}$ are the same. This is not always true but in general a metric is always symmetrical and satisfied:

$$g^{\alpha\gamma} g_{\gamma\beta} = g^{\alpha}_{\beta} = \delta^{\alpha}{}_{\beta} \qquad [A.12]$$

Thus, covariant and contravariant forms of a metric are the inverse of each other.

Metrics are used to bring an index down to the top and vice versa. Thus, for a vector **V**, we have:

$$V^{\alpha} = g^{\alpha\beta} V_{\beta} \qquad [A.13]$$

and:

$$V_{\alpha} = g_{\alpha\beta} V^{\beta} \qquad [A.14]$$

Another important aspect of metrics is their relationship to the interval. In fact, we have:

$$ds^2 = g_{\alpha\beta} dx^\alpha dx^\beta \qquad [A.15]$$

For example, in the case of the Minkowski space-time, in Cartesian coordinates, we have:

$$\begin{aligned} ds^2 &= \eta_{\alpha\beta} dx^\alpha dx^\beta \\ &= -dt^2 + dx^2 + dy^2 + dz^2 \end{aligned} \qquad [A.16]$$

It can happen that we want to derive metrics. For this purpose, *Christoffel symbols* $\Gamma$ are defined as linear transformations whose components are given by:

$$\Gamma^\alpha_{\beta\gamma} = \frac{1}{2} g^{\alpha\delta} \left( \frac{\partial}{\partial x^\gamma} g_{\delta\gamma} + \frac{\partial}{\partial x^\beta} g_{\delta\gamma} - \frac{\partial}{\partial x^\delta} g_{\beta\gamma} \right) \qquad [A.17]$$

Another use of the Minkowski metric in special relativity is to define a local Lorentz frame of reference.

## A.6. Divergence in space-time

To generalize the hydrodynamic approach in the relativistic context (see Chapter 4), it is necessary to define what a *divergence* means in a non-Euclidean space. We use the usual notation which consists of representing the derivatives of a tensor by a comma (,) followed by an index. Thus, for a tensor $\mathbf{T}$, we will write:

$$T^{\alpha\beta}_{,\gamma} := \frac{\partial}{\partial x^\gamma} T^{\alpha\beta} \text{ and } T_{\alpha\beta,\gamma} := \frac{\partial}{\partial x^\gamma} T_{\alpha\beta} \qquad [A.18]$$

With this convention, the traditional definition of the divergence of a vector $V^\alpha$ can be written as follows:

$$\nabla \cdot \mathbf{V} = V^\alpha_{,\alpha} = \frac{\partial V^0}{\partial x^0} + \frac{\partial V^1}{\partial x^1} + \frac{\partial V^2}{\partial x^2} + \frac{\partial V^3}{\partial x^3} \qquad [A.19]$$

and corresponds to a scalar.

Similarly, the covariant derivative of a four-vector $\mathbf{V}$ is a tensor $(1,1)$ whose components $(V^\alpha_{;\beta})$ are given by:

$$\left(\nabla_\beta \mathbf{V}\right)^\alpha := V^\alpha_{;\beta} := V^\alpha_{,\beta} + \Gamma^\alpha_{\mu\beta} V^\mu \qquad [A.20]$$

Similarly, the covariant derivative of a tensor $(2,0)$ $T^{\mu\nu}$ is a tensor $(2,1)$ whose components are given by:

$$T^{\mu\nu}_{;\lambda} = T^{\mu\nu}_{,\lambda} + T^{\alpha\nu}\Gamma^{\mu}_{\alpha\lambda} + T^{\mu\alpha}\Gamma^{\nu}_{\alpha\lambda} \qquad [A.21]$$

# References

Arbuzova, E.V., Dolgov, A.D., Reverberi, L. (2014). Jeans instability in classical and modified gravity. *Phys. Lett. B*, 739, 279.

Baym, G., Patil, S.P., Pethick, C.J. (2017). Damping of gravitational waves by matter. *Phys. Rev. D*, 96, 084033.

Bernal, A. and Guzmán, F.S. (2006). Scalar field dark matter: Nonspherical collapse and late-time behavior. *Phys. Rev. D*, 74, 103002.

Bessiri, A., Ourabah, K., Zerguini, T.H. (2021). Jeans instability in Eddington-inspired Born-Infeld (EiBI) gravity: A quantum approach. *Phys. Scr.*, 96, 125208.

Bhatnagar, P.L., Gross, E.P., Krook, M. (1954). A model for collision processes in gases. I. Small amplitude processes in charged and neutral one-component systems. *Phys. Rev.*, 94, 511.

Bricmont, J. (2021). La théorie de de Broglie-Bohm comme façon rationnelle de compléter la mécanique quantique. In *Ondes, matière et Univers*, Belaubre, G., Cohen-Tannoudji, G., Elbaz, C., Gondran, M., Mastrangelo, V., Schmets, J., Treuil, J.-P. (eds). EDP Sciences, Les Ulis.

Capozziello, S., De Laurentis, M., De Martino, I., Formisano, M., Odintsov, S.D. (2012). Jeans analysis of self-gravitating systems in gravity. *Phys. Rev. D*, 85, 044022.

Caroll, S.M. (1997). Lecture notes on general relativity [Online]. Available at: arXiv:gr-qc/9712019.

Chavanis, P.H. (2020). Quantum tunneling rate of dilute axion stars close to the maximum mass. *Phys. Rev. D*, 102, 083531.

Cohen, E.G.D. and Thirring, W. (1973). *The Boltzmann Equation Theory and Applications*. Springer-Verlag, Vienna.

Colomes, E., Zhan, Z., Oriols, X. (2015). Comparing Wigner, Husimi and Bohmian distributions: Which one is a true probability distribution in phase space? *J. Comp. Elec.*, 14, 894.

De Martino, I. and Capolupo, A. (2017). Kinetic theory of Jean instability in Eddington-inspired Born-Infeld gravity. *Eur. Phys. J. C*, 77, 715.

Del Zanna, L., Bucciantini, N., Londrillo, P. (2003). An efficient shock-capturing central-type scheme for multidimensional relativistic flows. II. Magnetohydrodynamics. *Astron. Astrophys.* 400, 397.

Eckart, C. (1940). The thermodynamics of irreversible processes. III. Relativistic theory of the simple fluid. *Phys. Rev.*, 58, 919.

Gomes, C. (2020). Jeans instability in non-minimal matter-curvature coupling gravity. *Eur. Phys. J. C*, 80, 633.

Gondran, M. and Gondran, A. (2014). Measurement in the de Broglie-Bohm interpretation: Double-slit, Stern-Gerlach, and EPR-B. *Phys. Res. Int.*, 605908.

Hammann, J. and Hasenkamp, J. (2013). A new life for sterile neutrinos: Resolving inconsistencies using hot dark matter. *JCAP*, 10, 044.

Hiley, B.J. (1999). Active information and teleportation. In *Epistemological and Experimental Perspectives on Quantum Physics*, Greenberger, D., Reiter, W.L., Zeilinger, A. (eds). Springer, Dordrecht.

Hiley, B.J. (2002). From the Heisenberg picture to Bohm: A new perspective on active information and its relation to Shannon information. In *Proc. Conf. Quantum Theory: Reconsideration of Foundations*, Khrennikov, A. (ed.). Vaxjö University Press.

Jeans, J.H. (1902). The stability of a spherical nebula. *Phil. Trans. Roy. Soc. A.*, 199, 312.

Kremer, G.M. and André, R. (2016). Analysis of instability of systems composed by dark and baryonic matter. *Int. J. Mod. Phys. D*, 25, 1650012.

Landau, L.D. (1946). On the vibrations of the electronic plasma. *J. Phys. USSR*, 10, 25.

Landau, L.D. and Lifshitz, E.M. (1959). *Fluid Mechanics*. Pergamon Press, Oxford.

Laplace, P.-S. (1814). *Essai philosophique sur les probabilités*. Courcier, Paris.

Ling, F.-S., Nezri, E., Athanassoula, E., Teyssier, R. (2010). Dark matter direct detection signals inferred from a cosmological N-body simulation with baryons. *JCAP*, 2, 012.

Lovell, M.R., Frenk, C.S., Eke, V.R., Jenkins, A., Gao, L., Theuns, T. (2014). The properties of warm dark matter haloes. *MNRAS*, 439, 300.

Madelung, E. (1927). Quantentheorie in hydrodynamischer form. *Z. Physik*, 40, 322.

Magaña, J. and Matos, T. (2012). A brief review of the scalar field dark matter model. *J. Phys.: Conf. Series*, 378, 012012.

Mahfouf, A. (2016). Coefficients de transport dans des plasmas hors de l'équilibre. PhD Thesis, Université Blaise Pascal, Clermont-Ferrand.

Manfredi, G., Hervieux, P.-A., Hurst, J. (2019). Phase-space modeling of solid-state plasmas. *Rev. Mod. Phys.*, 3, 13.

Mendonça, J.T. (2019). Wave-kinetic approach to the Schrödinger-Newton equation. *New J. Phys.*, 21, 023004.

Mendonça, J.T. and Terças, H. (2013). *Physics of Ultra-Cold Matter.* Springer, New York.

Mendonça, J.T., Bingham, R., Shukla, P.K. (2005). A kinetic approach to Bose-Einstein condensates: Self-phase modulation and Bogoliubov oscillations. *JETP*, 101, 942.

Mohazzabi, P. and Shankar, S. (2018). Maxwell-Boltzmann distribution in solids. *J. Appl. Math. Phys.*, 6, 602.

Moretti, F., Bombacigno, F., Montani, G. (2020). Gravitational Landau damping for massive scalar modes. *Eur. Phys. J. C*, 80, 1203.

Ngô, C. and Ngô, H. (2008). *Physique statistique*, 3rd edition. Dunod, Paris.

Ourabah, K. (2020a). Jeans instability in dark matter halos. *Phys. Scr.*, 95, 055005.

Ourabah, K. (2020b). Linear dark matter density perturbations: A Wigner approach. *EPL (Europhysics Letters)*, 132, 19002.

Ourabah, K. (2021). Fingerprints of nonequilibrium stationary distributions in dispersion relations. *Sci. Rep.*, 11, 12103.

Pani, P., Delsate, T., Cardoso, V. (2012). Eddington-inspired Born-Infeld gravity: Phenomenology of nonlinear gravity-matter coupling. *Phys. Rev. D*, 85, 084020.

Papadopoulos, P. and Font, J.A. (1998). Relativistic hydrodynamics around black holes and horizon adapted coordinate systems. *Phys. Rev. D*, 58, 024005.

Raimbault, J.-L. and Smets, R. (2014). Modélisations fluides des plasmas. Master's course, École polytechnique, Paris [Online]. Available at: https://www.lpp.polytechnique.fr/IMG/pdf/cours_2013.pdf.

Rezolla, L. and Zanotti, O. (2013). *Relativistic Hydrodynamics.* Oxford University Press.

Rothstein, J. (1974). Loschmidt's and Zermelo's paradoxes do not exist. *Found. Phys.*, 4, 83.

Savoini, P. (2010). Introduction à la physique des plasmas. Master's course, École polytechnique, Paris [Online]. Available at: https://www.lpp.polytechnique.fr/IMG/pdf/Cours_Intro_Plasma_M1_WEB.pdf.

Steckline, V.S. (1983). Zermelo, Boltzmann, and the recurrence paradox. *Am. J. Phys.*, 51, 894.

Trigger, S.A., Ershkovich, A.I., van Heijst, G.J.F., Schram, P.P.J.M. (2004). Kinetic theory of Jeans instability. *Phys. Rev. E*, 69, 066403.

Tsiklauri, D. (1998). Jeans instability of interstellar gas clouds in the background of weakly interacting massive particles. *Astrophys. J.*, 507, 226.

Vereshchagin, G.V. and Aksenov, A.G. (2017). *Relativistic Kinetic Theory with Applications in Astrophysics and Cosmology*. Cambridge University Press.

Vlasov, A.A. (1938). The vibrational properties of an electron gas. *JETP*, 8, 291.

Wigner, E.P. (1932). On the quantum correction for thermodynamic equilibrium. *Phys. Rev.*, 40, 749.

Würtz, P., Langen, T., Gericke, T., Koglbauer, A., Ott, H. (2009). Experimental demonstration of single-site addressability in a two-dimensional optical lattice. *Phys. Rev. Lett.*, 103(8), 080404.

Zel'dovich, Y.B. (1977). Hydrodynamics of the Universe. *Ann. Rev. Fluid Mech.*, 9, 215.

# Index

Other titles from

ISTE

in

Fluid Mechanics

## 2022

PRUD'HOMME Roger, VINCENT Stéphane
*Fluid Mechanics at Interfaces 1: Methods and Diversity*
*Fluid Mechanics at Interfaces 2: Case Studies and Instabilities*

## 2020

SAD CHEMLOUL Nord-Eddine
*Dimensional Analysis and Similarity in Fluid Mechanics*

## 2019

CALTAGIRONE Jean-Paul
*Discrete Mechanics: Concepts and Applications*

## 2018

ABABOU Rachid
*Capillary Flows in Heterogeneous and Random Porous Media 1*

## 2017

TARDU Sedat
*Wall Turbulence Control*

## 2015

CALTAGIRONE Jean-Paul
*Discrete Mechanics*

FOHR Jean-Paul
*Heat and Moisture Transfer between Human Body and Environment*

## 2014

PRUD'HOMME Roger
*Flows and Chemical Reactions in Heterogeneous Mixtures*
*Flows and Chemical Reactions in an Electromagnetic Field*

TARDU Sedat
*Transport and Coherent Structures in Wall Turbulence*

## 2013

BORGHI Roland, ANSELMET Fabien
*Turbulent Multiphase Flows with Heat and Mass Transfer*

DÉLERY Jean
*Three-dimensional Separated Flow Topology: Critical Points, Separation Lines and Vortical Structures*

PRUD'HOMME Roger
*Flows and Chemical Reactions in Homogeneous Mixtures*

## 2012

BOUTIER Alain
*Laser Velocimetry in Fluid Mechanics*

BOUTIER Alain
*Laser Metrology in Fluid Mechanics: Granulometry, Temperature and Concentration Measurements*

PRUD'HOMME Roger
*Flows and Chemical Reactions*

## 2011

MORY Mathieu
*Fluid Mechanics for Chemical Engineering*

TARDU Sedat
*Statistical Approach to Wall Turbulence*

## 2010

COLIN Stéphane
*Microfluidics*

DÉLERY Jean
*Handbook of Compressible Aerodynamics*

GUINOT Vincent
*Waves Propagation in Fluids – 2nd edition*

SOULI Mhamed, BENSON J. David
*Arbitrary Lagrangian Eulerian and Fluid-structure Interaction*

## 2009

FAVRE-MARINET Michel, TARDU Sedat
*Convective Heat Transfer*

## 2008

LALAUZE René
*Physical Chemistry of Solid-Gas Interfaces*

PEUBE Jean-Louis
*Fundamentals of Fluid Mechanics and Transport Phenomena*

SCHIESTEL Roland
*Modeling and Simulation of Turbulent Flows*

Printed and bound by CPI Group (UK) Ltd, Croydon, CR0 4YY
13/09/2026
14961041-0001